Natural Computing Series

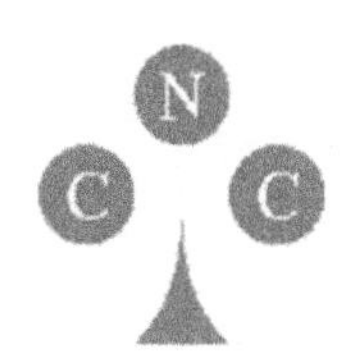

Springer
Berlin
Heidelberg
New York
Barcelona
Hong Kong
London
Milan
Paris
Singapore
Tokyo

William M. Spears

Evolutionary Algorithms

The Role of Mutation and Recombination

With 55 Figures and 23 Tables

Author

Dr. William M. Spears

AI Center - Code 5514
Naval Research Laboratory
4555 Overlook Avenue, S.W.
Washington, DC 20375, USA

spears@genetic-algorithms.com

Series Editors

G. Rozenberg (Managing Editor)
Th. Bäck, A.E. Eiben, J.N. Kok, H.P. Spaink

Leiden Center for Natural Computing
Leiden University
Niels Bohrweg 1
2333 CA Leiden, The Netherlands
rozenber@cs.leidenuniv.nl

Library of Congress Cataloging-in-Publication Data

Spears, William M., 1962–
Evolutionary algorithms: the role of mutation and recombination/William M. Spears.
p.cm. – (Natural computing series)
Includes bibliographical references and index.

1. Evolutionary programming (Computer science) 2. Computer algorithms. I. Title.
II.
Series.

QA76.618. S64 2000
005.1–dc21 00-037327

ACM Computing Classification (1998): F.1.1, F.2.2, I.2.8, I.2.6, J.3

ISBN 978-3-642-08624-3

Springer-Verlag is a company in the BertelsmannSpringer publishing group.

Printed in Germany

Cover Design: Künkel + Lopka, Werbeagentur, Heidelberg

For Diana

Preface

Despite decades of work in evolutionary algorithms (EAs), much uncertainty remains about the advantages and disadvantages of using recombination or mutation. This book provides a theoretical and empirical study of recombination and mutation in EAs, in order to better characterize the roles of these operators.

The main theme of the book is as follows. First, a static, component-wise analysis of recombination and mutation is performed. This analysis highlights some of the strengths and weaknesses of both operators. For example, the analysis suggests that increasing the number of peaks in a fitness landscape (i.e., its multimodality) can have a highly deleterious effect on an EA with recombination. This suggestion is confirmed via a dynamic Markov chain analysis of an EA on very small problems. The dynamic analysis also suggests that the relative heights of the peaks can influence the utility of recombination. Finally, these results are empirically confirmed on real problems through the use of a novel multimodality problem generator that produces random problems with a controllable amount of multimodality. When all peaks have equal heights, increasing the number of peaks has an increasingly deleterious effect on the performance of an EA with recombination. However, gradually lowering the heights of the suboptimal peaks is beneficial to the performance of recombination. Interestingly, the EA with mutation (and no recombination) is almost completely unaffected by the number of peaks or their heights.

As well as following the main theme, the book also takes occasional excursions into related theoretical areas, in order to unify the existing theoretical techniques more closely. Using a static analysis, a No Free Lunch theorem is proven for recombination, demonstrating a tight relationship between the disruptive and constructive aspects of recombination. An intriguing relationship is also demonstrated between uniform recombination and mutation when the cardinality of the representation is two and there is maximum population diversity. The book also shows a close relationship between the static analyses and a dynamic analysis of a population undergoing recombination and/or mutation by demonstrating that the more disruptive an operator is (a static concept), the faster the population approaches an equilibrium distribution (a dynamic concept).

Finally, the book introduces new techniques for studying EAs. First, it introduces a model of an EA with selection and mutation that involves iterating equations of motion. It then defines a class of functions under which this model can be naturally aggregated. This produces an identical model with far fewer equations of motion, allowing the model to be applied to realistic fitness functions. Finally, a novel aggregation algorithm is created to automatically aggregate a more complex Markov model of an EA that includes selection, mutation, and recombination. Since this aggregation algorithm works for general Markov chain models, it has a potential scope well beyond that examined in this book.

A note on the format and intended audience of this book is in order. It was written with the intention that it will typically be used as a reference book. As a consequence, there is more redundancy in the writing than would be the case for a pure textbook. This redundancy will be useful for any reader who wishes to jump immediately to particular chapters of interest.

Although the intended audience includes computer scientists and mathematicians interested in evolutionary algorithms, the chapters concerning recombination, mutation, and speciation should also be of interest to evolutionary biologists, population geneticists, and epidemiologists. The chapters on test-problem generators should be of interest to practitioners interested in evaluating and comparing different search (and optimization) algorithms in general. And finally, the Markov chain aggregation algorithm should be of interest to researchers in the operations research community.

As with any project of this scope, a large number of people contributed to the completion of this book. I thank Diana Gordon for not letting me quit when the going got tough. Her many comments also contributed significantly to the book – she is responsible for suggesting how to evaluate my Markov chain aggregation algorithm. I thank Vic Anand for finding an important mathematical error in an early draft. I also thank Kenneth De Jong for his extremely valuable guidance during the course of this work. I thank Thomas Bäck and Zbigniew Michalewicz for urging me to write this manuscript. This work would not have been completed without the kind support of Alan Meyrowitz and John Grefenstette at the Artificial Intelligence Center of the Naval Research Laboratory, who allowed me the time to develop many of the ideas in this book. Naturally, this work has been influenced by many other researchers – I especially thank Lashon Booker, Larry Eshelman, David Fogel, David Goldberg, Heinz Mühlenbein, Erik van Nimwegen, Terry Jones, David Schaffer, Gilbert Syswerda, and Michael Vose for their insights (although they may not agree with my interpretations at times). Mitch Potter helped enormously with the LaTeX2e. Finally, thanks to my parents for always encouraging my natural curiosity.

For those of you who are interested in corresponding with me, I can be reached at spears@genetic-algorithms.com.

Washington, DC, May 2000 William M. Spears

Table of Contents

Notation

Notation Used Throughout the Book

A, B, X, Y	General random variables
C	Cardinality of the alphabet
$E[X]$	Expected value of random variable X
H_k	Hyperplane of order k
K	Set of k defining positions in H_k
$\|K\|$	Cardinality of a set K
L	Length of strings
$L_1, ..., L_{k-1}$	Defining lengths of H_k
P	Population size
$P(X = x)$	Probability that random variable X has value x
P_0	Probability of swapping alleles in uniform recombination
$P_{\text{eq}}(i)$	Probability that two alleles at locus (position) i are equal
$PS(K)$	Power set of a set K
S, S_i, S_j, S_h	Various strings composed of L alleles
a_i	Allele at locus (position) i
$f(i), f_i$	Fitness of individual i
${p_i}^{(t)}$	Proportion of some item i at time t
μ	Mutation rate
χ	Recombination rate
λ	Number of offspring produced
$\mathcal{A}$	Alphabet of the individuals in the population
$\mathcal{P}$	Number of peaks in a search space

Notation Used for Schema Theory

$E_{\text{s}}[B_k]$	Expected number of offspring in H_k after survival
$E_{\text{c}}[B_k]$	Expected number of offspring in H_k after construction
$P_{\text{d}}(H_k)$	Probability of disrupting the hyperplane H_k
$P_{\text{s}}(H_k)$	Probability that hyperplane H_k will survive
$P_{\text{c}}(H_k)$	Probability of constructing the hyperplane H_k
$\mathit{\Gamma}$	Constructive advantage of recombination over mutation
$\mathcal{R}$	Set of all recombination events
$\mathcal{S}$	Set of all situations under which construction can occur

Notation Used for Markov Chains

EWT_J	Expected waiting time until event J occurs
I, J	Sets of states
N	Number of states in the Markov chain
$Q(i,j)$	Probability of transitioning from state i to j in one step
$Q^n(i,j)$	Probability of transitioning from state i to j in n steps
S_t	Markov chain random variable at time t
Z	Nix and Vose matrix of states
$m_{i,j}$	Mean passage time from state i to state j
$p_{i,j}$	Probability of transitioning from state i to j in one step
${p_{i,j}}^{(n)}$	Probability of transitioning from state i to j in n steps
${p_i}^{(t)}$	Probability of being in state i at time t
π_i	Steady-state probability of state i
$\mathcal{M}$	Nix and Vose mutation and recombination operator
$\mathcal{F}$	Nix and Vose fitness function operator

Notation Used for the Aggregation Algorithm

A, B, T, X, Y	Useful matrices
$Q \equiv Q_{\mathrm{u}}$	Uncompressed (nonaggregated) Markov chain
Q_{c}	Compressed (aggregated) Markov chain
$Error_{i,j}$	Error introduced by aggregating states i and j
$\alpha_{i,j}$	Row similarity of states i and j
$\beta_{i,j}$	Column similarity of states i and j

Part I

Setting the Stage

1. Introduction

1.1 Evolutionary Algorithms

Evolutionary computation uses computational models of evolutionary processes as key elements in the design and implementation of computer-based problem solving systems. A variety of evolutionary computational models have been proposed and studied – all of which are referred to as evolutionary algorithms (EAs). EAs share a common conceptual base of simulating the evolution of individual structures via processes of selection and perturbation. These processes depend on the perceived performance (fitness) of the individual structures as defined by an environment.

More precisely, evolutionary algorithms maintain a population of structures that evolve according to rules of selection and other operators, such as recombination and mutation. Each individual in the population is evaluated, receiving a measure of its fitness in the environment. Selection focuses attention on high-fitness individuals, thus exploiting the available fitness information. Recombination and mutation perturb those individuals, providing general heuristics for exploration. Although simplistic from a biologist's viewpoint, these algorithms are sufficiently complex to provide robust and powerful adaptive search mechanisms.

Figure 1.1 outlines a typical evolutionary algorithm (EA). A population of P individual structures is initialized and then evolved from generation t to generation $t+1$ by repeated applications of fitness evaluation, selection, recombination, and mutation. The population size P is generally constant in an evolutionary algorithm, although there is no a priori reason (other than convenience) to make this assumption.

An evolutionary algorithm typically initializes its population randomly, although domain-specific knowledge can also be used to bias the search. Evaluation measures the fitness of each individual according to its worth in some environment. Fitness evaluation may be as simple as computing a mathematical function or as complex as running an elaborate simulation. Selection is often performed in two steps, parent selection and survival. Parent selection decides who becomes parents and how many children the parents have; higher-fitness individuals are more likely to be parents and have more children. Children are created via recombination, which exchanges information between the parents, and mutation, which further perturbs the children. The

```
procedure EA;
t = 0; /* Initial Generation */
initialize_population(t);
evaluate(t);
until (done) {
        t = t + 1; /* Next Generation */
        select_parents(t);
        recombine(t);
        mutate(t);
        evaluate(t);
        select_survivors(t);
}
```

Fig. 1.1. The outline of an evolutionary algorithm

children are then evaluated. Finally, the survival step decides who survives in the population.

The origins of evolutionary algorithms can be traced to at least the 1950s (Fraser 1957; Box 1957). However, the three most historically significant methodologies are "evolutionary programming" (Fogel et al. 1966), "evolution strategies" (Rechenberg 1973; Schwefel 1981), and "genetic algorithms" (Holland 1975). These methodologies in turn have inspired the development of additional evolutionary algorithms such as "classifier systems" (Holland 1986), the LS systems (Smith 1983), "adaptive operator" systems (Davis 1989), GENITOR (Whitley 1989), SAMUEL (Grefenstette 1989), "genetic programming" (Koza et al. 1999), "messy genetic algorithms" (Goldberg 1991), and the CHC approach (Eshelman and Schaffer 1991). Excellent introductions to evolutionary algorithms and their application to real-world problems can be found in Goldberg (1989), Dasgupta and Michalewicz (1997), Michalewicz (1999), and Michalewicz (2000).

1.1.1 A Simple Example

Let us illustrate an EA with a simple example. Suppose an automotive manufacturer wishes to design a new engine and fuel system in order to maximize performance, reliability, and gas mileage, while minimizing emissions. Let us further suppose that an engine simulation unit can test various simulated engines and return a single value indicating the fitness score of the engine. However, the number of possible engines is large and there is insufficient time to test them all. How would one attack such a problem with an EA?

First, we define each individual to represent a specific engine. For example, suppose the cubic inch displacement (CID), fuel system, number of valves, cylinders, and presence of turbo-charging are all engine variables. The initialization step would create an initial population of possible engines. For the sake of simplicity, let us assume a (very small) population of size four. An example initial population is shown in Table 1.1.

Table 1.1. An initial sample population of an EA, before the individuals are evaluated

Individual	CID	Fuel System	Turbo	Valves	Cylinders
1	350	4 Barrels	Yes	16	8
2	250	Mech. Inject.	No	12	6
3	150	Elect. Inject.	Yes	12	4
4	200	2 Barrels	No	8	4

Table 1.2. An initial sample population of an EA, after the individuals are evaluated

Individual	CID	Fuel System	Turbo	Valves	Cylinders	Score
1	350	4 Barrels	Yes	16	8	50
2	250	Mech. Inject.	No	12	6	100
3	150	Elect. Inject.	Yes	12	4	300
4	200	2 Barrels	No	8	4	150

We now evaluate each individual with the engine simulator. Each individual receives a fitness score (the higher the better – see Table 1.2). Parent selection decides who has children and how many to have. For example, we could decide that the third individual deserves two children because it is so much better (i.e., has higher fitness) than the other individuals. Children are created through recombination and mutation. As mentioned above, recombination exchanges information between individuals, while mutation perturbs individuals, thereby increasing diversity. For example, recombination of the third and fourth individuals could produce the two children in Table 1.3.

Table 1.3. Recombination is performed on the third and fourth individuals.

Individual	CID	Fuel System	Turbo	Valves	Cylinders
$3'$	200	Elect. Inject.	Yes	8	4
$4'$	150	2 Barrels	No	12	4

Note that the children are composed only of elements from the two parents. Further note that the number of cylinders must be four, because the third and fourth individuals both had four cylinders. Mutation might further perturb these children, yielding the individuals in Table 1.4.

Table 1.4. Mutation is performed on the third and fourth individuals.

Individual	CID	Fuel System	Turbo	Valves	Cylinders
$3'$	250	Elect. Inject.	Yes	8	4
$4'$	150	2 Barrels	No	12	6

Table 1.5. The resulting children are now evaluated.

Individual	CID	Fuel System	Turbo	Valves	Cylinders	Score
3′	250	Elect. Inject.	Yes	8	4	250
4′	150	2 Barrels	No	12	6	350

We now evaluate the children (see Table 1.5). Finally we decide who will survive. In our constant population size example, which is typical of most EAs, we need to select four individuals to survive. How this is accomplished varies considerably in different EAs. If, for example, only the best individuals survive, our population would become as shown in Table 1.6. This cycle of evaluation, selection, recombination, mutation, and survival continues until some termination criterion is met.

Table 1.6. The next generation of an EA, after the individuals have been recombined, mutated, and re-evaluated

Individual	CID	Fuel System	Turbo	Valves	Cylinders	Score
3	150	Elect. Inject.	Yes	12	4	300
4	200	2 Barrels	No	8	4	150
3′	250	Elect. Inject.	Yes	8	4	250
4′	150	2 Barrels	No	12	6	350

This simple example serves to illustrate the flavor of an evolutionary algorithm. It is important to point out that although the basic conceptual framework of all EAs is similar, their particular implementations differ in many details. For example, there are a wide variety of selection mechanisms. The representation of individuals ranges from binary strings to real-valued vectors, Lisp expressions (Koza 1992), and neural networks. Finally, the relative importance of mutation and recombination differs widely across different methodologies. We describe the three most historically significant methodologies in the remainder of this section.

1.1.2 Evolutionary Programming

Evolutionary programming (EP), developed by Fogel et al. (1966), traditionally has used representations that are tailored to the problem domain. For example, in real-valued optimization problems, the individuals within the population are real-valued vectors. Similarly, ordered lists are used for traveling salesman problems, and graphs for applications with finite-state machines. EP is often used as an optimizer, although it arose from the desire to generate machine intelligence.

After initialization, all P individuals are selected to be parents, and then are mutated, producing P children. These children are evaluated and P survivors are chosen from the $2P$ individuals, using a probabilistic tournament selection. The best individual always survives, ensuring that once an optimum is found it cannot be lost. This is referred to as "elitism." The form of mutation is based on the representation used, and is often adaptive. For example, when using a real-valued vector, each variable within an individual may have an adaptive mutation rate that is normally distributed with a zero expectation. Recombination is not generally performed since the forms of mutation used are quite flexible and can produce perturbations similar to recombination, if desired.

The theoretical foundations for this algorithm stem from a proof of the global convergence (with probability 1) for EP (Fogel 1992). This result is derived by defining a Markov chain over the discrete state space that is obtained using the numbers represented on a digital computer. By combining all possible populations that contain the grid point having fitness closest to the true global optimum, an absorbing state is defined in which the process will ultimately be trapped, due to elitism.

1.1.3 Evolution Strategies

Evolution strategies (ESs) were developed by Rechenberg (1973), using selection, mutation, and a population of size one. Schwefel (1981) introduced recombination and populations with more than one individual, and provided a nice comparison of ESs with more traditional optimization techniques. Due to initial interest in hydrodynamic optimization problems, evolution strategies typically use real-valued, vector representations.

After initialization and evaluation, individuals are selected uniformly randomly to be parents. In the standard recombinative ES, pairs of parents produce children via recombination, and the children are further perturbed via mutation. The number of children created is greater than the number of parents P. Survival is deterministic and is implemented in one of two methods. The first method allows the P best children to survive, and replaces the parents with these children. This is referred to as (P, λ) survival, where P corresponds to the population size and λ refers to the number of children created. The second method is referred to as $(P + \lambda)$ survival, which allows the P best children *and* parents to survive. The $(P + \lambda)$ method is elitist, while the (P, λ) method is not.[1] Like EP, considerable effort has focused on adapting mutation as the algorithm runs by allowing each variable within an individual to have an adaptive mutation rate that is normally distributed with a zero expectation. Unlike EP, however, recombination does play an important role in evolution strategies, especially in adapting mutation.

Most of the theory for ESs is concerned with convergence velocity, i.e., trying to maximize the rate at which the ES converges to the optimum.

[1] It is more traditional to refer to these methods as (μ, λ) and $(\mu + \lambda)$, but this conflicts with the common use of μ for the mutation rate in EAs.

There are also proofs of global convergence (with probability 1) for the elitist $(P + \lambda)$ ES, but not for the (P, λ) ES. Bäck and Schwefel (1993) provide a nice overview of the theory.

1.1.4 Genetic Algorithms

Genetic algorithms (GAs), developed by Holland (1975), have traditionally used a more domain-independent representation, namely, binary strings. However, many recent applications of GAs have focused on other representations, such as graphs (neural networks), Lisp expressions, ordered lists, and real-valued vectors.

After initialization, parents are selected according to a probabilistic function based on relative fitness, referred to as "fitness-proportional selection." With this selection method the average fitness of the population is monitored. Those individuals that have higher than average fitness produce (on the average) more than one child, while those that have less than average fitness produce (on the average) less than one child. This is normalized appropriately to produce P children, which are created via recombination from the P parents. The P children are then mutated and replace the P parents in the population. This form of selection is not elitist and can be considered to be a (P, P) selection strategy.

The theoretical foundation of GAs is considerably different from that used in ES or EP. Holland (1975) concentrates on "schemata," which are sets of individuals. Holland likens these schemata to the random variables associated with K-armed bandit problems, and argues that the GA maximizes accumulated payoff by optimizing the allocation of trials to those random variables.

It is interesting to note that the relative emphasis on mutation and recombination in a GA is opposite to that in EP. Historically, in a GA, mutation is considered to be a background operator that flips alleles with some small probability. Recombination is considered to be the primary search operator. Despite this emphasis on recombination, interest in mutation has increased recently, partly due to the influence of the ES and EP communities. Schaffer and Eshelman (1991) have experimentally shown that mutation is a powerful search operator in its own right, while still maintaining the usefulness of recombination in certain situations.

1.1.5 Summary

As one can see, the three most historically significant EA methodologies vary considerably with respect to representation, selection, population management (e.g., how many children are created), and their use of recombination and mutation.[2] This book will focus primarily on recombination and mutation. Note that the relative importance of recombination and mutation

[2] For a more extensive overview see Spears et al. (1993).

in particular EAs varies enormously. In EP the emphasis is on mutation, whereas with GAs the emphasis is on recombination. ESs make extensive use of both mutation and recombination. Despite decades of work with these algorithms, however, there still remains a lot of uncertainty as to when it is good or bad to use recombination or mutation. The goal of this book is to provide a theoretical and empirical study of recombination and mutation in EAs, in order to better characterize the roles of these operators.

Almost all of the prior theoretical and empirical analyses of recombination have been performed in the GA community, and have assumed fixed-length, linear representations over finite-cardinality alphabets. Such representations include the binary-string representation often used in GAs, the ordered-list representation in EP, and even the real-valued, vector representation in ESs (since the values are represented in a digital computer). We will continue to make these assumptions throughout this book, since they hold in most applications of EAs. It is possible that the analyses in this book could be extended to variable-length and nonlinear representations, however, this will not be attempted here.

This book will make use of piecewise component analyses (of recombination and mutation in isolation) as well as full analyses of the complete dynamics of selection, recombination, and mutation. It will be shown that the component analyses provide useful insights into the dynamics of the complete analyses. Finally, the book will confirm those insights by use of a novel experimental methodology that uses a problem generator to create random problems within a well-defined class.

Before delving fully into the book, however, it is important to summarize the characterizations of recombination and mutation that have appeared in the literature, discuss related issues, summarize the goals of the book, and provide an outline for the remaining chapters.

1.2 Overview of Related Work

This section provides an overview of recombination and mutation theory. First, however, some useful terminology must be defined. An EA operates on a population of P strings, which are generally fixed length. The strings are of length L and are often referred to as "chromosomes" with L "genes." The position of each gene is its "locus." Each gene can take on one of C values, which are often referred to as "alleles." C can be thought of as the cardinality of the alphabet of the strings. Thus, there are C^L possible strings. This representation includes the real-valued representations used by the EP and ES methodologies (since the real values are represented in a digital computer); however, it maps most naturally to the discrete binary-string representations used in GAs. For this reason we borrow extensively from the terminology of the GA community.

Schemata (Holland 1975) represent sets of strings by using an additional symbol in the alphabet: #. For example, let us consider the schema AB##, defined over a fixed-length chromosome (individual) of four genes, where each gene can take on one of 26 alleles {A, ..., Z}. The # is defined to be a "don't care" (i.e., wildcard) symbol, and schema AB## represents all chromosomes that have an A for their first allele and a B for their second. Since each of the # symbols can be filled in with any one of the 26 alleles, this schema represents 26^2 chromosomes.

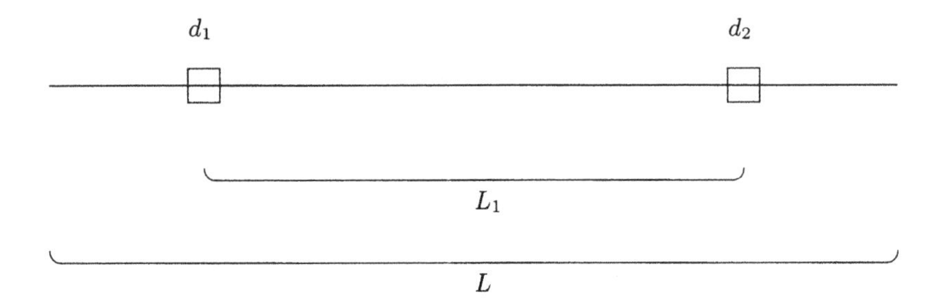

Fig. 1.2. A second-order hyperplane H_2

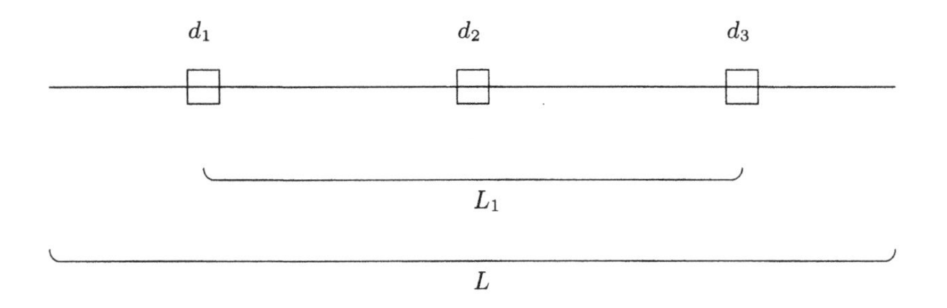

Fig. 1.3. A third-order hyperplane H_3

Schemata are also referred to as "hyperplanes" and "building blocks" in the literature (Goldberg 1987). Hyperplanes are described by "defining positions" d_i, that indicate where the non-# symbols occur. The "order" k is the number of defining positions (non-# symbols) in the hyperplane, and thus the hyperplane of order k is often designated as H_k. Also, since there are k defining positions, they are labeled d_1 through d_k. The "defining length" L_1 of a hyperplane is the distance between the outermost non-# symbols (d_1 and d_k). For example the schema ABC# has order three and defining length two, while the schema A##D has order two and defining length three. As further examples, consider Fig. 1.2 and Fig. 1.3, which pictorially depict second-order and third-order hyperplanes. In both figures the length of the individuals is L, and the defining length of the hyperplanes is L_1. The second-order hyperplane H_2 has two defining positions d_1 and d_2, whereas the third-order hyperplane has three defining positions d_1, d_2, and d_3. The small squares represent the alleles at those defining positions.

1.2.1 Recombination

Although Holland (1975) was not the first to suggest recombination in an evolutionary algorithm (EA) (e.g., see Fraser 1957 and Fogel et al. 1966), he was the first to place theoretical emphasis on this operator. According to Holland, an adaptive system must persistently test and incorporate structural properties associated with better performance. The object, of course, is to find *new* structures which have a high probability of improving performance significantly.

Holland concentrated on schemata, which provide a basis for associating combinations of attributes with potential for improving current performance. Suppose every chromosome has a well-defined fitness value (also called "utility" or "payoff"). Now suppose there is a population of P individuals, p of which are members of the schema AB##. The "observed average fitness" of that schema is the average fitness of those p individuals in that schema. It is important to note that these individuals will also be members of other schemata; thus the population of P individuals contains instances of a large number of schemata (all of which have some observed fitness). Holland (1975) stated that a good heuristic is to generate new instances of those schemata whose observed fitness is higher than the average fitness of the whole population, since instances of those schemata are likely to exhibit superior performance.

For example, suppose the schema AB## does in fact have a high observed fitness. The heuristic states that new samples (instances) of that schema should be generated. Selection (reproduction) does not produce *new* samples – but recombination or mutation can. The key aspect of recombination is that if one recombines two individuals that start with AB, their offspring must also start with AB. Thus one can retain what appears to be the promising building block AB##, yet continue to test that building block in new contexts. Mutation will not necessarily preserve the AB alleles.

Holland (1975) provided one of the earliest analyses of a recombination operator, called "one-point" recombination. Suppose there are two parents: AACC and ACAC. Randomly select one point at which to separate ("cut") both parents.[3] For example, suppose they are cut in the middle: AA|CC

[3] Since the individuals have a fixed length, the cut-point will necessarily be the same for both parents.

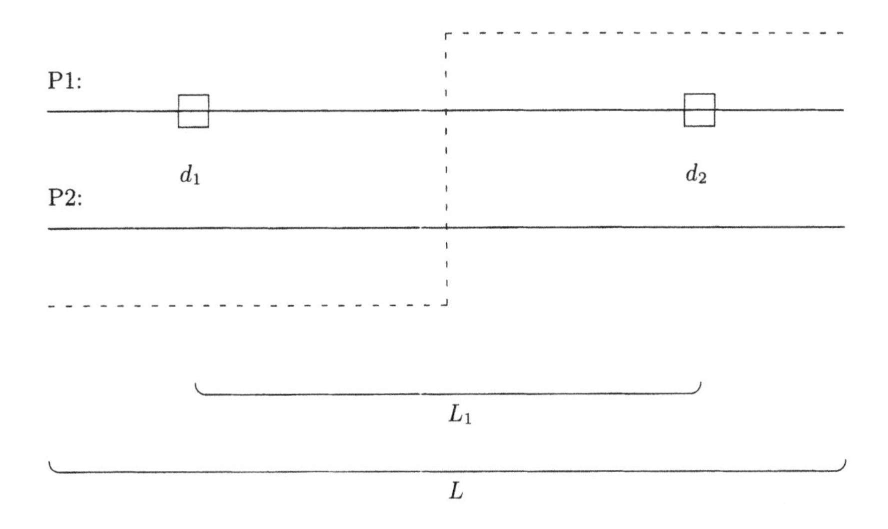

Fig. 1.4. How one-point recombination can disrupt a second-order hyperplane

and AC|AC. The offspring are created by swapping the tail (or head) portions to yield AAAC and ACCC. Holland analyzed one-point recombination by examining the probability that various schemata will be disrupted when undergoing recombination. A schema is disrupted if neither offspring is in that schema. For example, consider the two schemata AA## and A##A. Each schema can be disrupted only if the "cut-point" falls between its two As. This is much more likely to occur with the schema A##A than with AA##. In fact, the probability of disrupting either schema is proportional to the distance between the As. Thus, one-point recombination has the bias that it is much more likely to disrupt "long" schemata (those with a large defining length) than "short" schemata (those with a small defining length). Figure 1.4 provides a pictorial example. The two parents are labeled P1 and P2. Suppose P1 is a member of a particular second-order hyperplane H_2, which has defining positions d_1 and d_2, while P2 is some other arbitrary individual. Then H_2 can be disrupted if the cut-point falls between the two defining positions (see the dashed line, which represents the cut-point in one-point recombination). This becomes more likely as the distance between those defining positions (the defining length) increases.

De Jong (1975) extended this analysis to include so-called "n-point" recombination. In n-point recombination n cut-points are randomly selected and the genetic material between cut-points is swapped. For example, with two-point recombination, suppose the two parents AAAA and BBAB are cut as follows: A|AA|A and B|BA|B. Then the two offspring are ABAA and BAAB. De Jong noted that two-point (or n-point where n is even) recombination is less likely to disrupt "long" schemata than one-point (or n-point where n is odd) recombination.[4]

Syswerda (1989) introduced a new form of recombination called "uniform" recombination. Uniform recombination does not use "cut-points" but instead creates offspring by deciding, for each allele of one parent, whether to swap that allele with the corresponding allele in the other parent. That decision is made using a coin-flip (i.e., the swap is made with probability denoted as P_0). Syswerda compared the probability of schema disruption for one-point, two-point, and uniform recombination.[5] Interestingly, while uniform recombination is somewhat more disruptive of schemata than one-point and two-point, it does not have a length bias (i.e., the defining length of a schema does not affect the probability of disruption). Also, Syswerda showed that the more disruptive nature of uniform recombination can be viewed in another way – it is more likely to "construct" instances of new higher-order schemata from lower-order schemata than one-point and two-point recombination. Chapters 3–8 of this book extend this earlier work, by providing a common framework upon which to compare all n-point recombination and P_0 uniform recombination operators with mutation, on hyperplanes of arbitrary order, under varying conditions of population homogeneity.

Eshelman et al. (1989) considered other characterizations of recombination. They introduced two biases, the positional and distributional bias. A recombination operator has positional bias to the extent that the creation of any new schema by recombining existing schemata is dependent upon the location of the alleles in the chromosome. This is similar to the length bias introduced above. A recombination operator has distributional bias to the extent that the amount of material that is exchanged is not uniformly distributed. For example, one-point recombination has high positional (length) bias. However, it has no distributional bias, since one-point recombination will exchange anywhere from one to L alleles uniformly with probability $1/L$. Two-point recombination has slightly lower positional bias and still no distributional bias. Uniform recombination has no positional bias but high distributional bias because the amount of material exchanged is binomially distributed. Chapter 9 re-examines and extends this work by providing a comparison with mutation, under varying conditions of population homogeneity.

All of the analyses mentioned thus far are "static" in the sense that they do not attempt to model the time evolution of an EA. A more dynamic characterization is Geiringer's Theorem (Geiringer 1944), which describes the equilibrium distribution of an arbitrary population that is repeatedly undergoing recombination, but no selection or mutation. To understand Geiringer's Theorem, consider a population of ten strings of length four. Five of the strings

[4] Moon and Bui (1994) independently performed a similar analysis that emphasized "clusters" of defining positions within schemata, as opposed to lengths.

[5] Syswerda (1989) only considered uniform recombination where $P_0 = 0.5$.

are AAAA while the other five are BBBB. If these strings are recombined repeatedly, eventually $2^4 = 16$ strings will become equally likely in the population. In equilibrium the probability of a particular string will approach the product of the probabilities of the individual alleles – thus asserting a condition of independence between alleles. Booker (1992) suggests that the rate at which the population approaches equilibrium is the significant distinguishing characterization of different recombination operators. Chapter 10 extends this earlier work by attempting to perform that characterization for different recombination operators. Chapter 10 also provides a related analysis for mutation.

1.2.2 Mutation

As one can see, recombination theory has centered around the GA model of evolutionary computation. The theory for mutation has centered around the ES model. Rechenberg (1973) investigated the (1+1) ES on corridor and sphere models of fitness and computed the expected rates of convergence, given the standard deviation ("step size") of the normally distributed mutation operator. The standard deviations were then optimized to yield the fastest rates of convergence. It was then possible to compute the probability that a given mutation would be "successful" (have better fitness than the parent). Since this was close to 1/5 for both the sphere and corridor models, the "1/5 rule" was formed (Rechenberg 1973): "The ratio of successful mutations to all mutations should be 1/5. If it is greater than 1/5 increase the standard deviation; if it is smaller, decrease the standard deviation."

Schwefel (1981) extended the analysis to population sizes greater than one, and computed the expected progress of the population average for the $(P + \lambda)$ and (P, λ) ES. This allowed Schwefel to estimate that having a λ/P ratio of roughly six yields the best compromise between the computational effort of having a large number of offspring and the convergence rate.

It might be possible to extend these analyses to include recombination, by computing the "success probability" (i.e., the probability that offspring will be better than parents) for various recombination operators on various problem classes. However, this will not be pursued in this book.

1.2.3 Differences between Recombination and Mutation

Although prior work has helped to distinguish between different recombination operators in GAs and to select good step sizes for mutation in ESs, it has not helped to illustrate the differences between recombination and mutation. However, the notion of building blocks has led Fogel (1995) to hypothesize that recombination will perform poorly for most naturally evolved systems, because (so he claims) they are extensively pleiotropic (a gene may influence multiple traits) and highly polygenic (a trait may be influenced by multiple genes). Since such systems will not have many high-fitness building blocks for recombination to exploit, Fogel argues that mutation will be superior for these systems.

The biological concepts of pleiotropy and polygeny are related to another concept called "epistasis." A system has low (high) epistasis if the optimal allele for any locus depends on a small (large) number of alleles at other loci. Systems with independent loci (the optimal allele for each locus can be decided independently of the alleles at the other loci) have no epistasis. A small amount of evidence exists that recombination is most useful for medium epistasis problems, but not for high epistasis problems (Schaffer and Eshelman 1991; Davidor 1990). This is consistent with Fogel's hypothesis. Thus the concept of epistasis would appear to be useful for exploring the differences between recombination and mutation. This is explored further in Chaps. 2 and 14.

Another useful concept is the "operator landscapes" of Jones (1995). In this model the fitness landscape is treated from an operator point of view. Landscapes are commonly described in Hamming space, which is an ideal view for mutation, since the mutation of a parent yields a child that is nearby in Hamming space. However, Hamming distance is not necessarily useful when considering recombination, since the recombination of two parents can yield children arbitrarily far in Hamming space. Jones argues that in order to study any operator one needs to consider how far apart points are in operator space, not just Hamming space. For example, if one considers standard notions of "multimodality," these notions are generally embedded firmly in Euclidean or Hamming space. However, when seen from the point of view of recombination, the space may look entirely different. This indicates that multimodality might also be a useful concept for exploring the differences between recombination and mutation. This is explored further in Chaps. 2, 12, and 14.

1.3 Issues and Goals

As pointed out above, one of the central unanswered issues in EAs is a useful characterization of the strengths and weaknesses of recombination and mutation. What are the similarities and differences between mutation and recombination? When will recombination or mutation help or hurt performance? In order to answer these questions, this book will use both theoretical and empirical approaches.

1.3.1 Issues in Theoretical Approaches

Rather than pick a particular theoretical tool for addressing these questions, this book will take a more eclectic approach. There is a continuum of levels for modeling complex systems, ranging from the simple to the complex.

Simple theoretical models are generally easy to analyze, but are approximations to the real system. As the approximations are improved, the models generally become more complex, increasing the analytical burden. This book will examine the full continuum – simple, component-size models (schema and limiting distribution models of recombination and mutation), intermediate models of selection and mutation, and complete models of EAs using Markov chains. The point is that, as with a good set of Craftsman tools, it is often necessary to have many different levels of modeling in order to address different questions.

One criticism of simple models is that they often include too few details of a system, resulting in a model that is totally nonpredictive in nature. This criticism is often levied against one of the most common theoretical techniques for studying recombination – so-called "schema" theories, in which the disruptive and constructive aspects of recombination on hyperplanes (schemata) are compared. On the other hand, very complete models are also problematic due to their analytical complexity. For example, Markov chain models of EAs can have enormous numbers of states, raising a criticism concerning the usefulness of Markov chain theories for EAs.

This book will address both criticisms. It will show that Markov theories can in fact provide quite useful insights – the behavior of an EA on small (computationally tractable) problems can in fact be observed in larger problems (Chaps. 12 and 14). Furthermore it is possible to provide automatic tools for simplifying these complete models to make them far more computationally manageable (Chap. 13). Finally, it will be shown that the results from the simple schema theories (Chap. 8) provide the inspiration for the experiments performed with the Markov model of an EA (Chap. 12), indicating that a theory need not be totally predictive to be useful.

1.3.2 Issues in Empirical Approaches

As mentioned above, the theories developed in this book either make simplifying assumptions, or (when simplifying assumptions are not made) are analytically tractable only on small problems. Thus it is necessary to test these theories empirically in order to ensure that the theoretically derived conclusions and hypotheses hold with real evolutionary algorithms on large, realistic problems. One weakness of standard empirical studies in which search algorithms are compared is that their results may not generalize beyond the test problems used. A classic example of this is a study in which a new algorithm is carefully tuned to the point that it outperforms some existing algorithms on a few ad hoc problems (e.g., the De Jong 1975 test suite). The results of such studies typically have only weak predictive value regarding relative performance on new problems. Thus the central issue here is in how to perform better empirical studies, in order to better characterize when EAs will perform well or poorly.

There are two ways to strengthen the results obtained from empirical studies. The first is to remove the opportunity to hand-tune algorithms to a particular problem or set of ad hoc problems. This can be done by using "test-problem generators," which produce random problems from within a well-specified class of problems. Having problem generators allows one to report results over a randomly generated set of problems that have well-controlled characteristics, rather than a few hand-chosen, ad hoc examples. Thus, by increasing the number of randomly generated problems, the predictive power of the results for the problem class as a whole has increased. An advantage of problem generators is that in most cases they are quite easy to parameterize, allowing one to design controlled experiments in which one or more properties of a class of problems can be varied systematically to study the effects on particular search algorithms. Chapter 14 gives empirical results using a test-problem generator inspired by the theoretical results obtained with the complete Markov model from Chap. 12.

On a related issue, it is common practice to run EAs to some fixed termination criteria, and then to report the results only after termination. However, this ignores the dynamic aspects of an EA, and can lead to overly general conclusions. For example, as we will see, conclusions can often turn out to be surprisingly dependent on the termination criteria, often reversing if a different cutoff is used. Thus, a second way to improve empirical methodology is to include results throughout the running of an algorithm. We will generally show results over the whole running time of an EA.

1.4 Outline

This book has a central theme with occasional excursions into related areas. The central theme proceeds as follows. First, the book will introduce a novel test-problem generator based on Boolean satisfiability problems, which will serve to further clarify the relative roles of mutation and recombination on epistatic and multimodal problems (Chap. 2). After this the book will start to delve into the theoretical aspects of recombination, by generalizing the traditional static schema theory for recombination. The purpose of the theory is to compute the disruptive aspects that n-point recombination and P_0 uniform recombination have on kth-order hyperplanes H_k (Chap. 3). Then the theory will be generalized further to include the constructive aspects of recombination (Chap. 4). At this point a schema theory for mutation can be derived, which also takes into account the disruptive (Chap. 6) and constructive (Chap. 7) aspects of mutation. Population homogeneity and arbitrary cardinality alphabets will be taken into account in a natural fashion. Mutation and recombination can thus be fairly compared, and various general hypotheses will be made concerning the relative aspects of recombination and mutation. This comparison occurs in Chap. 8.

The book then introduces a complete dynamic model of an EA, which uses Markov chains (Chap. 12). The experiments performed with the Markov chain model are inspired by the insights gained from the static schema theories from Chaps. 3–8. The results of the Markov chain approach yield further hypotheses concerning the role of mutation and recombination on simple test functions.

Finally, the book confirms those hypotheses by examining the performance of an actual EA on real functions. A test-problem generator is created (Chap. 14), motivated by the results from the Markov chain approach. The results of the experiments validate the results from the schema and Markov chain theories, completing the central theme.

However, this book also takes a number of interesting excursions, which often help to clarify the roles of mutation and recombination in EAs. First, we use the mathematical technique for handling population homogeneity to generalize prior static analyses of the distributional and positional biases of recombination, and provide a comparison with mutation (Chap. 9). Then we examine simple dynamic theories concerning the distribution of populations undergoing recombination and mutation (in the limit of large time) and connect these theories with the prior static schema theories (Chap. 10). We also provide a simple dynamic model that includes selection and mutation, and illustrate that for certain classes of problems an expectation model of a simple EA (without recombination) can be built that handles realistically large problems (Chap. 11). These excursions shed more insight into the roles of recombination and mutation, and provide new theoretical tools for examining simple EAs.

Finally, the book (in perhaps its most important excursion), gives an automatic algorithm for simplifying Markov chain theories in general (Chap. 13). Although motivated by a study of EAs, the algorithm will work on arbitrary Markov chains that are derived from other complex systems, and hence has a scope well beyond that examined in this book.

2. Background

2.1 No Free Lunch

As mentioned in the previous chapter, one of the central unanswered issues in EAs is a useful characterization of the strengths and weaknesses of recombination and mutation. Naturally, this task would be made easier if in fact it could be proven that either mutation or recombination were in some sense always superior to the other (e.g., an EA without recombination will always outperform an EA with recombination). After all, as Schwefel (1995) states: "A practitioner would much rather manage with just one strategy, which can solve all the practically occurring problems for as small a total cost as possible."

Unfortunately, there is a growing consensus that such "universal algorithms" do not appear to be possible, whether the algorithms are EAs or not. For example, Booker (1992) states that "there is no 'universal' bias that facilitates induction across all problem domains." Similarly, Hart and Belew (1991) conclude that "theoretical and experimental analyses of the GA which do not specify the class of functions being optimized can make few claims regarding the efficiency of the genetic algorithm for an arbitrary fitness function."

Most recently, Wolpert and Macready (1997) provided a solid theoretical framework for comparing the performance of optimization algorithms over all possible problems, concluding that: "Roughly speaking, we show that for both static and time dependent optimization problems, the average performance of any pair of algorithms across all possible problems is exactly identical." This has important ramifications (Wolpert and Macready 1997):

> In particular, if an algorithm performs better than random search on some class of problems then it must perform worse than random search on the remaining problems. Thus comparisons reporting the performance of a particular algorithm with particular parameter settings on a few sample problems are of limited utility. While such results do indicate behavior on the narrow range of problems considered, one should be very wary of trying to generalize those results to other problems.

The theorems of Wolpert and Macready are referred to as "No Free Lunch" theorems, because any increase in performance of an algorithm on certain problems must be offset by a decrease in performance on other problems. What this indicates is that studies in which an EA is carefully tuned to outperform some other algorithm on a few ad hoc problems have only weak predictive value regarding relative performance on new problems. Instead, the focus of research should be on identifying classes of EAs (or algorithms in general) that perform well and poorly on classes of problems under various measures of performance. The goal is to identify problem and algorithm characteristics that yield useful predictive theories concerning performance. From this perspective an EA theory should provide ways of measuring degrees of hardness of a particular problem. It should provide insight into the effects that changes in representation, operators, etc. have on hardness, and for a given EA make predictions about the kinds of problems with which it will have difficulties.

Said another way, the difficulty of a particular problem is strongly correlated to how well matched (in terms of some performance measure) the features of a particular EA are to the characteristics of that problem. High degrees of consonance correspond to our informal notion of an "EA-easy" situation, and significant dissonance results in "EA-hard" situations. As EA engineers we can and do frequently increase the consonance of a particular situation by changing representations, operators, etc. This view is consistent with the No Free Lunch theorems, which mathematically describe the dissonance and consonance as a formal dot product between an algorithm vector and a problem vector.

Thus one way to strengthen the results obtained from empirical studies is to generalize performance results over a broad class of problems with well-defined characteristics. This can be done by using "test-problem generators" that produce random problems from within a well-specified problem class. The advantage of test-problem generators is that it allows the user to methodically control important characteristics of problems.

As mentioned in the previous chapter, two important characteristics of problems are their "epistasis" and their "multimodality." The goal then is to develop a mechanism for methodically creating problems where these characteristics (as well as others) can be easily controlled. This chapter illustrates how Boolean satisfiability problems provide one such mechanism, since the epistasis and multimodality are well-defined and are easily controlled.

Once the Boolean satisfiability problem generator is defined, the chapter proceeds to illustrate how such problems can highlight the differences between mutation and recombination. This will involve the use of a simple "speciating" EA that simultaneously evolves separate subpopulations. For a constant population size, the presence of more subpopulations means that each subpopulation has fewer individuals. We then show, on a particular multimodal problem, that if the speciating EA performs better (from a function optimization point of view) when there are more subpopulations, recombination is adding very little value to the search process. On the other hand (as shown with an epistatic problem), if increasing the number of subpopulations has a detrimental effect on performance, recombination is very important for the search process. These experiments are not meant to be fully rigorous; they in fact serve only to motivate the theoretical and empirical studies on multimodality and epistasis conducted in the remainder of the book.

2.2 A Boolean Satisfiability Problem Generator

Boolean satisfiability (SAT) refers to the task of finding a truth assignment that makes an arbitrary Boolean expression true. The Boolean expression is composed of conjunctions, disjunctions, and negations of Boolean variables, where the Boolean variables can be *true* or *false*. For example, suppose there are two Boolean variables a and b. Then the Boolean expression $a \wedge \bar{b}$ is true if and only if the Boolean variable a is true and b is false ($\bar{b}$ is the logical negation of b).

In order to apply EAs to any particular problem, we need to select an appropriate representation for the solution space and define an external evaluation function which assigns utility to candidate solutions. Both components are critical to the success or failure of the EAs on the problem of interest. Fortunately, SAT has a simple string representation, namely, binary strings of length L. The ith bit represents the truth value of the ith Boolean variable of the L Boolean variables present in the Boolean expression.

Somewhat more thought must be given to selecting an evaluation function. The simplest and most natural function assigns a fitness of 1 to a candidate solution (string) if the Boolean values specified by that string result in the Boolean expression evaluating to *true*, and 0 otherwise. A moment's thought, however, suggests that for problems of interest the fitness function would be 0 almost everywhere and would not support the formation of useful schemata. Thus it is necessary to reward partial solutions to SAT problems, in order to provide intermediate feedback to the EA.

One approach to providing intermediate feedback is to transform a given Boolean expression into conjunctive normal form (CNF) and define the fitness to be the total number of top level conjuncts which evaluate to true.[1] While this makes some intuitive sense, one cannot in general perform such transformations in polynomial time without introducing a large number of additional Boolean variables which, in turn, combinatorially increase the size of the search space.

An alternative approach is to assign fitness to individual subexpressions in the original expression and combine them in some way to generate a total

[1] CNF refers to Boolean expressions that are a conjunction of clauses, each clause being a disjunction of negated or nonnegated Boolean variables.

fitness value. In this context the most natural approach is to define the fitness of *true* to be 1, the value of *false* to be 0, and to define the fitness f of simple expressions as follows:

$$\begin{aligned} f(\bar{e}) &= 1 - f(e) \\ f(e_1 \wedge e_2 \wedge \cdots \wedge e_n) &= MIN(f(e_1), f(e_2), \cdots f(e_n)) \\ f(e_1 \vee e_2 \vee \cdots \vee e_n) &= MAX(f(e_1), f(e_2), \cdots f(e_n)) \end{aligned}$$

Since any Boolean expression can be broken down (parsed) into these basic elements, one has a systematic mechanism for assigning fitness. Unfortunately, as the astute reader has probably already noticed, this mechanism is no better than the original one since it still only assigns fitness values of 0 and 1 to both individual subexpressions and the entire expression.

However, a minor change to this mechanism can generate differential fitnesses, namely:

$$f(e_1 \wedge e_2 \wedge \cdots \wedge e_n) = AVE(f(e_1), f(e_2), \cdots f(e_n))$$

where AVE returns the average value of its parameters. This suggestion was made first by Smith (1979) and intuitively justified by arguing that this would reward "more nearly true" conjunctions. So, for example, solutions to the Boolean expression $(a \wedge (a \vee \bar{b}))$ would be assigned fitnesses as follows:

Table 2.1. Fitness f for assignments to $a \wedge (a \vee \bar{b})$

a	b	Fitness f
0	0	$AVE(0, MAX(0, (1-0))) = 0.5$
0	1	$AVE(0, MAX(0, (1-1))) = 0.0$
1	0	$AVE(1, MAX(1, (1-0))) = 1.0$
1	1	$AVE(1, MAX(1, (1-1))) = 1.0$

Notice that both of the correct solutions (lines 3 and 4) are assigned a fitness of 1 and, of the incorrect solutions (lines 1 and 2), line 1 gets higher fitness because it satisfied half of the conjunction.

This approach was used successfully by Smith and was initially adopted in our experiments. However, there were a number of features of this fitness function that left us uncomfortable and which led to a more careful examination of it.

The first and fairly obvious property of using AVE to evaluate conjunctions is that the fitness function is not invariant under standard Boolean equivalency transformations. For example, it violates the associativity law:

$$f((a \wedge b) \wedge c) \neq f(a \wedge (b \wedge c))$$

since

$$AVE(AVE(a,b),c) \neq AVE(a,AVE(b,c))$$

We have attempted to construct alternative differential fitness functions which have the ideal property of fitness invariance and have had no success. However, one could argue that a weaker form of invariance might be adequate for use with EAs, namely, *truth invariance.* By that we mean that the fitness function should assign the same value (1) to all correct solutions of the given Boolean expression, and should map all incorrect solutions into a set of lower values ($0 \leq value < 1$). Since Boolean transformations do not occur *while* the EAs are searching for solutions, the actual values assigned non-solutions would seem to be of much less importance than the fact that they are useful as a differential fitness to support the construction of partial solutions.

Unfortunately, the proposed fitness function does not even guarantee this second and weaker property of truth invariance, as the following shows:

$$a \vee b = \overline{(\overline{a} \wedge \overline{b})} \quad \text{by De Morgan}$$

However,

$$MAX(a,b) \neq 1 - \frac{((1-a)+(1-b))}{2}$$

as we see in the following table:

Table 2.2. Fitness f for assignments to $a \vee b = \overline{(\overline{a} \wedge \overline{b})}$

a	b	Left side	Right side
0	0	0.0	0.0
0	1	1.0	0.5
1	0	1.0	0.5
1	1	1.0	1.0

Notice that lines 2–4 are all solutions, but lines 2 and 3 are assigned a fitness of 1/2 after De Morgan's law has been applied. In general, it can be shown that, although the fitness does not assign the value of 1 to non-solutions, it frequently assigns values less than 1 to perfectly good solutions and can potentially give higher fitness to non-solutions!

A careful analysis, however, indicates that these problems *only* arise with expressions of the form $\overline{(a \wedge b \cdots)}$. This suggests a simple fix: preprocess each Boolean expression by systematically applying De Morgan's laws to remove such constructs. It also suggests another interesting opportunity. Constructs of the form $\overline{(a \vee b \cdots)}$ are computed correctly, but only take on 0/1 values. By using De Morgan's laws to convert these to conjunctions, we introduce additional differential fitness. Converting both forms is equivalent to reducing the scope of all negations to simple variables. Fortunately, unlike the conversion to CNF, this process has only linear complexity and can be done quickly and efficiently.

In summary, with the addition of this preprocessing step, we now have an effective fitness function for applying EAs to Boolean satisfiability problems. This fitness function has the following properties: 1) it assigns a fitness value of 1 if and only if the candidate solution is an actual solution; 2) it assigns values in the range $0 \leq value < 1$ to all non-solutions; 3) non-solutions receive differential fitness on the basis of how near their conjunctions are to being satisfied.

2.3 Using SAT to Create Multimodal and Epistatic Problems

The mapping in the previous section allows us to map any satisfiability problem into a function optimization problem amenable to an EA. One nice aspect of examining SAT problems is that it provides an easy vehicle for creating problems with certain known characteristics.

2.3.1 Using SAT to Create Multimodal Problems

For example, it is a straightforward process to create multimodal problems, which have multiple peaks. Consider the following expression of 30 Boolean variables:

$$\text{1Peak} \equiv (x_1 \ \wedge \ x_2 \ \wedge \ \cdots \ \wedge \ x_{30})$$

This conjunction is true if and only if all variables are true. Furthermore, given the mapping above, the fitness of the conjunction is simply the number of true variables (divided by 30). Thus, there is one "peak" in the problem, with fitness 1.0, and the more true variables there are the closer the fitness is to 1.0.

Now consider the following modification to the one-peak problem:

$$\text{2Peak} \equiv \text{1Peak} \ \vee \ (\overline{x_1} \ \wedge \ \overline{x_2} \ \wedge \ \cdots \ \wedge \ \overline{x_{30}})$$

By adding a disjunct at the highest level, we have created a problem with two peaks. One peak occurs where all variables are true (as before), but a new peak has been defined to occur where all variables are false (thus satisfying the second disjunct in the disjunction). Both peaks have fitness 1.0.

It is also possible to ensure that the fitness of the second peak is slightly less than 1.0, by making the second disjunct impossible to satisfy:

$$\text{2Peak} \equiv \text{1Peak} \ \vee \ (x_1 \ \wedge \ \overline{x_1} \ \wedge \ \overline{x_2} \ \wedge \ \cdots \ \wedge \ \overline{x_{30}})$$

Since it is impossible to have x_1 be true and false at the same time, the second disjunct cannot be satisfied. However, if the remaining 29 variables are false, the disjunct will have a high fitness just slightly less than 1.0. This is often referred to as a "false peak" problem, which has one optimal peak and one other peak that is almost optimal. The height of this false peak can be controlled (lowered) by simply adding more conflicting variables.

Clearly it is easy to create problems with even more peaks:

$$\begin{aligned}
\text{3Peak} &\equiv \text{2Peak} \lor (x_1 \land \overline{x_1} \land \overline{x_2} \land \cdots \land \overline{x_{15}} \land x_{16} \land \cdots \land x_{30}) \\
\text{4Peak} &\equiv \text{3Peak} \lor (x_1 \land \overline{x_1} \land x_2 \land \cdots \land x_{15} \land \overline{x_{16}} \land \cdots \land \overline{x_{30}}) \\
\text{5Peak} &\equiv \text{4Peak} \lor (x_1 \land \overline{x_1} \land x_2 \land \overline{x_3} \land x_4 \land \overline{x_5} \cdots \land \overline{x_{29}} \land x_{30}) \\
\text{6Peak} &\equiv \text{5Peak} \lor (x_1 \land \overline{x_1} \land \overline{x_2} \land x_3 \land \overline{x_4} \land x_5 \cdots \land x_{29} \land \overline{x_{30}})
\end{aligned}$$

For each of these problems the Boolean expression is true (and the fitness function is 1.0) if and only if all variables are true. However, each problem with $\mathcal{P}$ peaks has $\mathcal{P} - 1$ false peaks, with fitness slightly less than 1.0. The location of each false peak is determined by the disjunct that is associated with that peak. Thus we have shown that disjunctive SAT expressions are easily mapped to multimodal problems for an EA. The location and height of each peak is easily controlled by simply modifying the Boolean expression. We investigate this more thoroughly in Chap. 14.

2.3.2 Using SAT to Create Epistatic Problems

On the other hand, it turns out that epistatic problems can easily be created with conjunctive SAT expressions. A system has low (high) epistasis if the optimal allele for any locus depends on a small (large) number of alleles at other loci. Systems with independent loci have no epistasis. For such systems the optimal allele for each locus can be decided independently of the alleles at the other loci.

The biological concepts of pleiotropy and polygeny are related to epistasis. Pleiotropy refers to how many traits are influenced by a gene and polygeny refers to how many genes influence a trait. Roughly speaking, systems with low epistasis have low pleiotropy and low polygeny, while systems with high epistasis have high pleiotropy and/or high polygeny.

Consider the following simple conjunctive Boolean expression of four variables:

$$(a \lor b) \land (c \lor d)$$

Each conjunct can be considered to be a trait of the individual. The polygeny of the individual is the number of variables in each conjunct while the pleiotropy is the number of conjuncts each variable is in. In this case the polygeny and pleiotropy are low and the expression is easy to satisfy. However, if we consider a more complicated example, we can see how increasing pleiotropy and polygeny can increase the difficulty of the problem:

$$(a \lor b \lor \bar{c} \lor \bar{d}) \land (\bar{a} \lor \bar{b} \lor c \lor d) \land$$
$$(\bar{a} \lor \bar{b} \lor \bar{c} \lor d) \land (\bar{a} \lor b \lor c \lor \bar{d})$$

We can imagine a problem generator that randomly creates conjunctive expressions like the one above (and in fact we will discuss this further in Chap. 14). One difficulty with a problem generator of this type is that it is hard to know a priori if a solution actually exists. For the sake of illustration, we outline another approach here that can create epistatic problems of various sizes that always have one and only one solution. As with the multimodal problems, having only one solution makes the problem more difficult to solve. The key to this technique is to create specific Hamiltonian circuit problems, which are then converted into SAT expressions.

The Hamiltonian circuit (HC) problem consists of finding a tour through a directed graph that touches all nodes exactly once. Clearly, if a graph is fully connected, this is an easy task. However, as edges are removed the problem becomes much more difficult, and the general problem is known to be NP-Complete. Attempting to solve this problem directly with EAs raises many of the same representation issues as in the case of traveling salesman problems (De Jong 1985). However, it is possible to construct a polynomial-time transformation from HC problems to SAT problems (Garey and Johnson 1979).

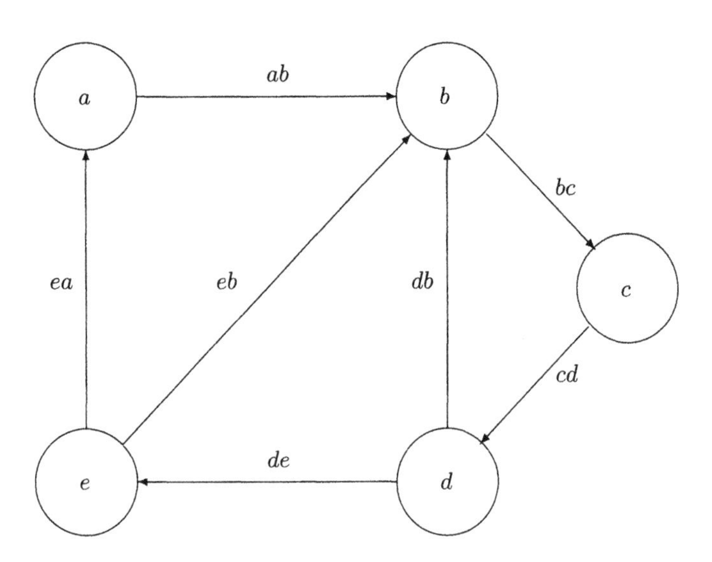

Fig. 2.1. An example Hamiltonian circuit problem with five nodes and seven edges

An example Hamiltonian circuit problem is given in Fig. 2.1. The example has five nodes, labeled $\{a, b, c, d, e\}$ and seven directed edges. There is only one tour through the edges: $\{ab, bc, cd, de, ea\}$. The definition of any HC problem implies that, for any solution, each node must have exactly one input edge and one output edge. If any tour violates this constraint, it cannot be a solution.[2] Therefore, the SAT problem equivalent to the HC problem is given by the boolean expression that is the conjunction of terms indicating valid edge combinations for each node. For the example shown in Fig. 2.1 the equivalent Boolean expression is:

$$\begin{gathered} ab \wedge bc \wedge cd \wedge de \wedge ea \wedge \\ ((db \wedge \overline{de}) \vee (\overline{db} \wedge de)) \wedge ((ea \wedge \overline{eb}) \vee (\overline{ea} \wedge eb)) \wedge \\ ((ab \wedge \overline{db} \wedge \overline{eb}) \vee (\overline{ab} \wedge db \wedge \overline{eb}) \vee (\overline{ab} \wedge \overline{db} \wedge eb)) \end{gathered}$$

As an example, consider node d. Node d has two output edges (labeled db and de), and one input edge cd. The output edge constraints are given by the exclusive-or $((db \wedge \overline{de}) \vee (\overline{db} \wedge de))$. The input edge constraint is given simply by cd. The assignments to the edge variables indicate which edges make up a tour, with a value of 1 indicating an edge is included and a value of 0 if it is not. The problem is somewhat epistatic, because some conjuncts are composed of several boolean variables and some of the boolean variables show up in multiple conjuncts. The transformation to the boolean expression is computed in polynomial time, and a solution to the HC problem exists if and only if the boolean expression is satisfiable.

To create epistatic problems with only one solution, we have defined the following family of HC problems. Consider a graph of N nodes, which are labeled using consecutive integers 1 through N. Suppose the first node has directed edges to all nodes with larger labels (except for the last node). The next $N-2$ nodes have directed edges to all nodes with larger labels (including the last node). The last node has a directed edge back to the first node. This yields $N(N-1)/2$ edges. A complete tour consists of following the node labels in increasing order, until you reach the last node. From the last node you travel back to the first. Because the edges are directed, it is clear that this is also the only legal tour. Increasing N increases the epistasis (see De Jong and Spears 1989 for more details on these problems). We refer to these problems as HCN (e.g., if $N = 8$ the problem is HC8).

In summary, we have shown how Boolean satisfiability problems provide a useful framework for methodically creating multimodal and epistatic problems. We will now show how such problems can illustrate the differences between mutation and recombination. This will involve the application of a simple "speciating" EA on the 6Peak problem (the multimodal problem with one true peak and five false peaks) and on the HC8 problem (which is

[2] In the general case, these constraints are necessary but not sufficient. However, for the cases considered here they are also sufficient.

reasonably epistatic). We show that the speciating EA performs better on the multimodal 6Peak problem as the number of subpopulations increases, and that recombination is adding very little value to the search process. On the other hand, on the epistatic HC8 problem, increasing the number of subpopulations in the speciating EA has a detrimental effect on performance, and recombination is very important for the search process. Before going into detail on these experiments, however, it is necessary to summarize what a speciating EA is, and how our particular implementation works.

2.4 Speciation

One of the attractive features of an EA is that it quickly concentrates effort in promising areas of a search space. Unfortunately, this feature is not always advantageous for many applications of interest. In such cases the attractive feature is described in a negative tone as "premature convergence." Although this phrase is subjective, it refers to the phenomenon in which the EA loses population diversity before some goal is met. Two typical cases are the optimization of multimodal functions and covering problems. In the former case the EA may concentrate effort on some suboptimal peak. In the latter case the problem is even more severe – the EA user wishes to simultaneously find a number of peaks, but the EA has lost all diversity and concentrates on one peak only.

Thus, one recurring theme in EA research is how to maintain population diversity. The mutation operator is often suggested as a solution. However, although high mutation rates certainly increase population diversity, they are often too destructive. In other words, the price of continuing exploration is that good solutions are often lost. The general consensus, then, is that diversity for the sake of diversity is not the issue. Rather it is the *appropriate* use of diversity, to explore new areas while not destroying the information already learned. One natural and appealing method of accomplishing this is by allowing the EA to *speciate* (i.e., evolve subpopulations). Each subpopulation can then explore a separate portion of the search space.

2.4.1 Restricted Mating and Sharing

As explained in Chap. 1, in a traditional generational EA P individuals reproduce according to their fitness (given by some objective function) and create a set of offspring via the application of genetic operators. The P parents are then replaced by P offspring to produce the next generation. One method for evolving species in a generational EA is through the use of *sharing* and *restricted mating* (see Mahfoud 1995 and Giordana and Neri 1995 for examples). With sharing, the similarity of individuals within the population is used to dynamically modify the fitness of those individuals. This dynamic modification is internal to the EA; thus what is changing is the EA's perception of the fitness function, not the fitness function itself. The intuition is simple – a peak in the space is treated as a resource that a species can exploit. Thus individuals near one peak (which are similar individuals that in essence form a species) have to share the resource of that peak. Overcrowding on one peak implies that a resource is overused, i.e., there are too many individuals of that species. In this case the perceived fitness of the peak goes down, reducing selective pressure in that area of the space. On the other hand, peaks with few individuals have their perceived fitness increased, increasing selective pressure in those areas. The net effect is to apportion individuals in rough proportion to the relative height of the peaks in the space.

Restricted mating also relies on the similarity metric, by restricting recombination (mating) to those individuals that are most similar. This helps prevent recombination from occurring between individuals of different species, which can produce children of poor fitness.

Although sharing and restricted mating are rather general ideas, the implementation of those ideas that has received the most attention is by Goldberg and Richardson (1987). In their implementation, sharing and restricted mating together work very well, creating stable species on many peaks within the search space. Furthermore higher peaks get more individuals than lower peaks, in rough proportion to the relative heights of the peaks, as we would expect with the sharing mechanism. Goldberg's implementation makes two assumptions. The first is the number of peaks in the space. The second is that those peaks are uniformly distributed throughout the space. However, no sensitivity study indicates how well these assumptions must be met. Finally, the implementation is also expensive. The similarity of each pair of individuals is measured, resulting in $O(P^2)$ similarity comparisons.

One possibility for reducing the complexity of the implementation is through sampling, as is done by Goldberg et al. (1992). However, it still relies on a similarity metric. An alternative to a similarity metric is to use *labels* to identify individuals. The motivation for this alternative comes from nature. For example, this author does not decide he is Portuguese because he is in some sense similar to a few of his cousins. Rather, he decides he is Portuguese because his ancestors were labeled as Portuguese. Although this does not mean that one genetically inherits such labels, they certainly can be culturally inherited. What if each EA individual has a label? Similarity then becomes simply a matter of seeing if two individuals have the same label. Of course, this implies that everyone with the same label is equally similar. However, results of Spears (1994) and Ryan (1995) indicate that the added precision of the similarity metric is often not needed.

Another difficulty with the use of a similarity metric to perform speciation is that it is hard to apply to EAs that have individuals with unusual representations. For example, suppose the individuals are Lisp expressions. Creating a meaningful similarity metric would be quite difficult. Thus, another advantage of labels is that it will work with any EA, regardless of the form that individuals take.

In summary, then, labels are an elegant substitute to a similarity metric. By implementing restricted mating and sharing with labels, the efficacy of restricted mating and sharing can be simultaneously accomplished in an efficient manner. Furthermore, labels can be used in any EA, regardless of the representation of individuals. A method for achieving speciation through the use of labels is outlined next.

2.4.2 A Simple Speciation Mechanism Using Labels

As mentioned above, we will attach a label to each EA individual. Labels can be represented in any arbitrary fashion, but for the sake of simplicity we assume they are *tag bits*, which are appended to every individual. Thus, l tag bits can be used to represent 2^l subpopulations. Each individual lies in the unique subpopulation described by its tag bits. Zero tag bits refers to the case of one subpopulation, i.e., a standard EA. Increasing the number of tag bits l increases the number of subpopulations, but only a reasonable number of subpopulations can be contained within any population. For example, with a population size of 100, five tags bits (32 subpopulations) implies that each initial subpopulation has a very small size of roughly three individuals. If more subpopulations are required, the population size should be increased.

We can now examine how sharing can be implemented with labels. Suppose that at every generation we count the number of individuals in each subpopulation. Then we normalize the observed fitness of each individual by dividing the fitness of each individual by the size of its subpopulation. Thus, a large subpopulation decreases the perceived fitness of a peak, allowing search to be focussed in other areas.

Sharing is easily described mathematically. Suppose the fitness of individual i is denoted as f_i. Further suppose that at a particular generation of the EA the P individuals are partitioned into k sets ($k \leq 2^l$), because there are k different values for the tag bits. We denote these sets as $\{s_0, \ldots, s_{k-1}\}$. The sets are numbered arbitrarily. Each individual belongs to one s_i and all individuals in a particular s_i have the same tag bit values. For example, suppose there is only one tag bit and that some individuals exist with a tag bit value 0 and that the remainder exist with tag bit value 1. Then we can (arbitrarily) assign the former set of individuals to s_0 and the latter set to s_1.

If we use $|\cdot|$ to denote the cardinality of the sets, then with sharing the perceived fitness, F_i, is a normalization of the objective fitness f_i:

$$F_i = \frac{f_i}{|s_j|}, i \in s_j$$

where $|s_j|$ is the size of the subpopulation that contains individual i.

To get an intuitive understanding of how this works, let us consider the following thought experiment. Suppose we have a simple function with two peaks, one peak twice as high as the other, and further suppose we allow one tag bit for each individual. Each tag bit is randomly initialized, so at the beginning of the EA run we have two subpopulations of roughly equal size. Due to random sampling both subpopulations could eventually settle on the higher peak, or both could settle on the lower peak. However, in some cases (again due to random sampling), each subpopulation will head towards different peaks. If we did not have fitness sharing, the individuals on the higher peak would always get more children than the individuals on the lower peak and eventually the subpopulation on the lower peak would vanish. However, with fitness sharing, the higher peak can support only twice as many individuals as can be supported on the lower peak (since it is only twice as high). In other words, the higher peak can support roughly 2/3 of the individuals, while the lower peak supports 1/3. To see this more clearly, suppose the population is of size 30 and that the fitness of the higher peak is 4, while the fitness of the lower peak is 2. Then, if we have 20 individuals at the high peak, their perceived fitness is 4/20. Meanwhile the remaining 10 individuals at the low peak have perceived fitness $2/10 = 4/20$. The fitness sharing mechanism has dynamically adjusted the perceived fitness so that the two peaks have the same perceived height. The result is that both subpopulations can survive in a stable fashion.

Now that sharing has been implemented with labels, all that remains is to show how restricted mating is implemented with labels. This is quite simple – recombination (mating) occurs only between individuals within the same subpopulation. To see why this is important, consider again the two-peak example described above.

Suppose that the higher peak is represented by the individual 111...111 while the lower peak is represented by the individual 000...000. Individuals with roughly 50% 1s and 0s are the lowest fitness strings, while individuals with mostly 1s or mostly 0s have high fitness. Mutation of any high-fitness individual on either peak will tend to keep the individual on that peak, driving it up or down the peak to a small degree. Recombination, however, produces quite different results, depending on the location of the parents. If the two parents are on the same peak, the offspring are also highly likely to be on that peak. However, if the two parents are on the two different peaks, the offspring are highly likely to be in the valley between the two peaks, where the fitness is low. Restricted mating prevents recombination between individuals on the two different peaks, preventing the construction of individuals lying in the valley.

The "simple subpopulation scheme" (SSS) is defined to be the algorithm that combines both the restricted mating and this simple sharing mechanism (see Fig. 2.2). SSS is a very effective and elegant speciation algorithm (see Spears 1994 for more details). The algorithm does not make the assumption

```
procedure SSS;
t = 0; /* Initial Generation */
initialize_population(t);
evaluate_with_sharing(t);
until (done) {
        t = t + 1; /* Next Generation */
        select_parents(t);
        recombine_within_subpopulations(t);
        mutate(t);
        evaluate_with_sharing(t);
        select_survivors(t);
}
```

Fig. 2.2. The outline of a speciating evolutionary algorithm

that the peaks are distributed in any particular fashion throughout the space. Since different subpopulations may lie on the same peak, the algorithm also does not make strong assumptions about the number of peaks in the space. One should be careful, though, to have more subpopulations than the number of peaks you wish to find. Both restricted mating and sharing can be executed in $O(P)$ time, due to the presence of the tag bits. Diversity is usefully maintained and the EA can divide attention among many areas of high fitness.[3]

2.4.3 Function Optimization Results

Now that the SSS algorithm has been described, it can be applied to some Boolean satisfiability problems defined earlier in this chapter. We will focus on both the 6Peak and HC8 problems, since they are nice simple representatives of multimodal and epistatic problems, respectively.

There are many ways to measure the performance of the SSS algorithm on these problems. For the sake of simplicity, we choose to measure the number of times the (unique) global optimum is found by the algorithm, given 100 independent runs on each problem. Each run is for 50,000 evaluations (i.e., 50,000 individuals have their fitness evaluated), and the population size ranges from 100 to 400 in increments of 100. The number of subpopulations ranges from 1 to 8 (recall that when there is one subpopulation that SSS is equivalent to a standard EA). Fitness-proportional selection is used (see Chap. 1) as well as recombination and mutation. Table 2.3 present the results of SSS on the HC8 problem.

Two observations can immediately be made from Table 2.3. The first is that for a given number of subpopulations, increasing the population size helps performance greatly. However, for a given population size, increasing

[3] SSS was inspired by research in artificial immune systems, where a peak can be considered to be an *antigen* and the evolved individuals are *antibodies* (Smith et al. 1992; Fukuda et al. 1999).

Table 2.3. Percentage of runs where SSS finds the optimum of HC8

	Population	Size		
Subpopulations	100	200	300	400
1	38%	69%	77%	95%
2	39%	69%	78%	84%
4	31%	67%	80%	94%
8	30%	62%	76%	90%

the number of subpopulations does not help performance (and often hinders performance). One possible reason for this performance has to do with premature convergence. As discussed in the introduction of Sect. 2.4, this refers to the phenomenon in which the EA loses population diversity too rapidly. When this happens the individuals are extremely similar to each other and recombination has little effect. One method for counteracting the loss of diversity is through the use of larger populations. Since the performance of SSS improves greatly with increased population size, this implies that recombination is a necessary component for successful search on this problem. To test this hypothesis we reran the experiments, but with recombination turned off. The results are shown in Table 2.4.

Table 2.4. Percentage of runs where SSS without recombination finds the optimum of HC8

	Population	Size		
Subpopulations	100	200	300	400
1	4%	6%	2%	3%
2	3%	1%	5%	4%
4	7%	9%	8%	4%
8	13%	11%	6%	5%

The results are quite striking and confirm the hypothesis. Without recombination SSS has much greater difficulty solving the problem. This is in line with the view that recombination can be useful for problems with medium epistasis (Schaffer and Eshelman 1991; Davidor 1990).

Table 2.5. Percentage of runs where SSS finds the optimum of 6Peak

	Population	Size		
Subpopulations	100	200	300	400
1	29%	42%	39%	50%
2	48%	57%	64%	65%
4	69%	72%	78%	80%
8	89%	87%	91%	93%

Let us now consider the performance of SSS on the multimodal 6Peak problem (see Table 2.5). The immediate observation is that increasing the number of subpopulations dramatically improves performance, whereas increasing the population size has only a modest effect on performance. This behavior is quite different from the behavior of SSS on the HC8 problem. Following the reasoning we used above, one possible hypothesis is that the loss of population diversity is not deleterious for this problem. This suggests that recombination is not nearly as important for the search process as it was for the HC8 problem. To test this hypothesis we again reran the experiments, but with recombination turned off. The results are shown in Table 2.6.

Table 2.6. Percentage of runs where SSS without recombination finds the optimum of 6Peak

	Population Size			
Subpopulations	100	200	300	400
1	25%	28%	20%	34%
2	40%	45%	48%	51%
4	54%	63%	58%	34%
8	53%	66%	40%	28%

The results confirm the hypothesis. Although the removal of recombination has a deleterious effect on performance the effect is not nearly as strong as it was with the HC8 problem. Thus it appears as if mutation (recombination) has a stronger (weaker) role in the successful search of 6Peak than in the search of HC8.

2.5 Summary

The No Free Lunch theorems of Wolpert and Macready (1997) indicate that it is impossible for any particular optimization or search algorithm to perform better than all other algorithms on all problems. In fact, it is guaranteed that if an algorithm is modified to perform better on some problem, it must therefore perform worse on some other problem. Thus, showing that an algorithm performs well on a small set of ad hoc problems is typically not extremely useful, since such results rarely indicate in any general fashion where the algorithm will perform well and poorly.

One way to strengthen the results obtained from empirical studies is to generalize performance results over a broad class of problems with well-defined characteristics. This can be done by using test-problem generators that produce problems from within a well-specified problem class. The advantage of test-problem generators is that it allows the user to methodically control important characteristics of problems.

This chapter has illustrated how Boolean satisfiability problems provide a simple and elegant framework for generating problems with well-defined and controllable characteristics. Examples were provided to show how disjunctive normal form (DNF) Boolean expressions map well to multimodal problems, where both the number and the height of the peaks are controlled by the syntax of the Boolean expressions. Similarly, conjunctive normal form (CNF) expressions are useful for describing epistatic problems, where both the pleiotropy and polygeny are controlled by the syntax of the Boolean expressions.

This chapter then provided a brief summary of speciation techniques for EAs and outlined a simple technique (called SSS) for converting a standard EA into a speciating EA. The technique relies on the concept of labels, which uniquely indicate to which species (subpopulation) an individual belongs. Labels can be attached to individuals regardless of their representation, and they provide an efficient means of performing fitness sharing and restricted mating. With fitness sharing an individual's perceived fitness depends on the objective fitness and the number of individuals in its species. This allows the EA to maintain species on different peaks in the space – higher peaks can hold somewhat more individuals than lower peaks. The restriction of mating (recombination) to members of a species helps to prevent poor fitness individuals.

We then show that on a problem with medium epistasis (HC8) recombination is very important for the search process. On the other hand, recombination is far less important for successful search on a simple multimodal problem (6Peak). These results hold even when SSS only has one subpopulation (and is hence equivalent to a standard generational EA).

Again, it should be pointed out that these results are not meant to be fully rigorous – they merely set the stage for the remainder of the work in this book. For example, it will be shown that multimodality is in fact an extremely valuable characteristic for illustrating the differences between mutation and recombination. Recall our earlier problem with two peaks at 111...111 and 000...000. Recombination between parents on different peaks is likely to produce offspring that lie in the valley between the two peaks, where the fitness is low. What if there are more than two peaks? It is reasonable to hypothesize that recombination could be even more deleterious, since the recombination of individuals on different peaks is even more likely to produce poor offspring, until the population has converged to one peak. It is these types of hypotheses and issues that drive the remainder of the book.

Part II

Static Theoretical Analyses

3. A Survival Schema Theory for Recombination

3.1 Introduction

The motivation for the original schema analysis of Holland (1975) was to compute the expected number of instances of hyperplanes at time $t+1$, given their number at time t. To use the notation of Goldberg (1987), let $m_t(H)$ be the number of individuals in hyperplane H at time t. Then let $f_t(H)$ be the observed average fitness of the hyperplane at time t and let $\bar{f}_t$ be the observed average fitness of the population at time t. Then, assuming fitness-proportional selection, the *expected* number of individuals in H at time $t+1$ is:

$$m_{t+1}(H) \ \geq \ m_t(H)\frac{f_t(H)}{\bar{f}_t}P_{\text{survival}}(H)$$

where $P_{\text{survival}}(H)$ is the probability that the hyperplane will not be disrupted by *either* mutation or recombination (i.e., it survives). The inequality refers to the fact that not only may a hyperplane H survive, it also may be *constructed* from other hyperplanes. See Holland (1975) or Goldberg (1987) for further discussions of this equation. The key point to be made here, however, is that computing the expected number of individuals in a hyperplane will yield a framework upon which to compare recombination and mutation.

That recombination is typically a two-parent operator while mutation is a one-parent operator presents some difficulties when trying to compare the two operators. The most important difficulty is that population homogeneity cannot be taken into account if mutation is treated as a one-parent operator (since population homogeneity is a relationship between multiple individuals in a population). This makes a comparison with recombination problematic, since recombination is dramatically affected by population homogeneity. The only apparent solution is to treat both operators as two-parent operators. Granted, mutation acts on each parent independently, but as long as that is taken into account, a fair comparison can be made. The framework taken in this book is to imagine two particular parents that undergo recombination or mutation (but not both). By computing the expected number of offspring that will exist in some hyperplane H after recombination or mutation, the two operators can be fairly compared.[1]

This chapter first computes the probability that a hyperplane H will survive recombination. It then uses this value to compute the expected number of offspring that will survive recombination. Chapter 4 then considers the *constructive* aspects of recombination by computing the probability that a hyperplane H can be constructed from other hyperplanes, using recombination. This value is then used to compute the expected number of offspring in H that will be constructed via recombination. Finally, Chapters 6–8 provide similar computations for mutation (that compute the expected number of offspring in H), allowing for a fair comparison between the two operators. Again, the key point is that it is the computation of the expected number of offspring in a hyperplane H that provides the common framework upon which to compare mutation and recombination.

3.2 Framework

Assume that individuals in an evolutionary algorithm are fixed-length strings whose characters are from a finite-cardinality alphabet. Let the length be L and the cardinality be C. For example, if the alphabet is the set of lower-case characters from the English alphabet, then $C = 26$ and two possible strings of length $L = 5$ are *hello* and *world*. For now we assume that there will be C^L possible strings.

Let a hyperplane of order k be denoted by H_k. H_k represents C^{L-k} possible strings, where the strings must match on the k defining positions of H_k. For example, $H_4 = aaaa\#$ is a fourth-order hyperplane that represents the 26 lower-case strings of length five that start with *aaaa*.

Let recombination be described by the random variable $\mathcal{R}$. For H_k there are 2^k possible recombination events: $0 \leq \mathcal{R} \leq 2^k - 1$. Each recombination event $\mathcal{R}$ can be represented by a bit mask of length k (i.e., the binary representation of $\mathcal{R}$), where a '1' at position j indicates that recombination swapped the alleles at position j between the two parents, and a '0' means that recombination did not swap the alleles at position j. All n-point recombination events and all parameterized uniform recombination events can be described with these bit masks. By definition, the probability of all recombination events $P(\mathcal{R})$ must sum to 1.

$$\sum_{\mathcal{R}} P(\mathcal{R}) = 1$$

[1] This framework does not appear to have obvious limitations; for example, it is simple to derive the one-parent analysis of mutation from the two-parent analysis, if one desires.

Assume that the following random experiment is being performed.[2] One is given two parents, and one parent is in the schema H_k, while the other parent is an arbitrary string (which may or may not be in the schema H_k). Figure 3.1 provides a pictorial example. The two parents are labeled P1 and P2. P1 is a member of a particular third-order hyperplane H_3, which has defining positions d_1, d_2, and d_3, and has defining length L_1. P2 is some other arbitrary individual.

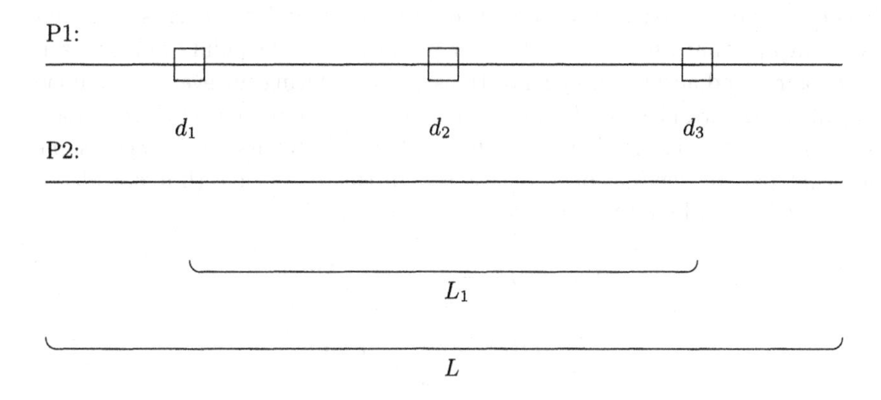

Fig. 3.1. The setup for the random experiment to be performed. P1 is a member of a third-order hyperplane, and P2 is an arbitrary string. Recombination is performed, producing two offspring.

The random experiment consists of performing recombination on these two parents, producing two children. Schema H_3 (or H_k in general) can either survive or be disrupted. A schema survives if either offspring is in H_3 (H_k), and it is disrupted if neither offspring is in H_3 (H_k). The probability of survival will be denoted as P_s, while the probability of disruption is denoted as P_d. Clearly,

$$P_s(H_k) = \sum_{\mathcal{R}} P(\mathcal{R})P(H_k \text{ survives} \mid \mathcal{R}) = \sum_{\mathcal{R}} P(\mathcal{R})P_s(H_k \mid \mathcal{R})$$

$$P_d(H_k) = \sum_{\mathcal{R}} P(\mathcal{R})P(H_k \text{ disrupted} \mid \mathcal{R}) = \sum_{\mathcal{R}} P(\mathcal{R})P_d(H_k \mid \mathcal{R})$$

Hyperplanes H_k will either survive or be disrupted, when subjected to recombination:

$$P_s(H_k) + P_d(H_k) = 1$$

[2] "A random experiment is simply an experiment in which the outcomes are non-deterministic, that is, probabilistic." (Stark and Woods 1986)

The remainder of this chapter concentrates on deriving (estimates of) the probability of disruption (or survival) of arbitrary order hyperplanes H_k under the action of n-point recombination and P_0 uniform recombination. These probabilities are then used to derive the expected number of offspring that will reside in H_k.

3.3 Survival Theory for n-point Recombination

With n-point recombination n cut-points determine the recombination event. It is convenient to partition these events into *even* and *odd* events. An even event means that an even number (or none) of the cut-points fall between *each* pair of adjacent defining positions of H_k. During an even event none (or all) of the k alleles in H_k are exchanged by recombination. In contrast, an odd event means that an odd number of the cut-points fall between *some* adjacent pair of defining positions of H_k. In this case some (but not all) of the k alleles will be exchanged by recombination. Thus,

$$P_{\text{even}}(H_k) = \sum_{\mathcal{R} \text{ even}} P(\mathcal{R})$$

$$P_{\text{odd}}(H_k) = \sum_{\mathcal{R} \text{ odd}} P(\mathcal{R})$$

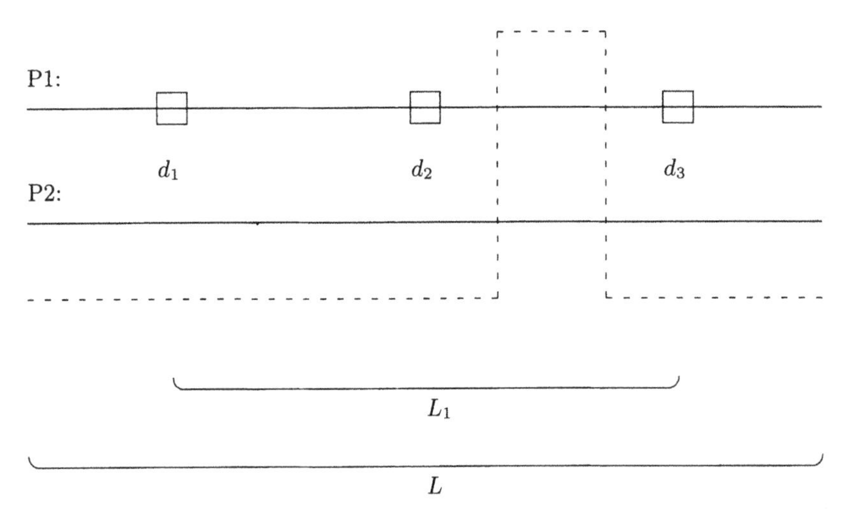

Fig. 3.2. An example of an even recombination event for two-point recombination

It is important to realize that a particular recombination event $\mathcal{R}$ will be even or odd only in relation to a particular hyperplane H_k. Figures 3.2 and 3.3 illustrate examples of even and odd recombination events, for two-point

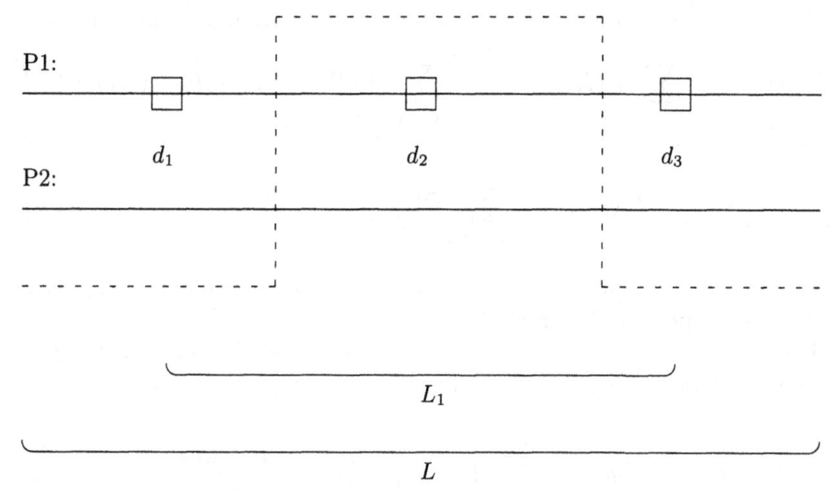

Fig. 3.3. An example of an odd recombination event for two-point recombination

recombination operating on a third-order hyperplane. Clearly, even and odd recombination events partition the set of all events, so their probabilities must sum to one:

$$P_{\text{even}}(H_k) + P_{\text{odd}}(H_k) = 1$$

Now one can write:

$$P_{\text{d}}(H_k) = \sum_{\mathcal{R}\ \text{even}} P(\mathcal{R})P_{\text{d}}(H_k \mid \mathcal{R}) + \sum_{\mathcal{R}\ \text{odd}} P(\mathcal{R})P_{\text{d}}(H_k \mid \mathcal{R}) \tag{3.1}$$

$$P_{\text{s}}(H_k) = \sum_{\mathcal{R}\ \text{even}} P(\mathcal{R})P_{\text{s}}(H_k \mid \mathcal{R}) + \sum_{\mathcal{R}\ \text{odd}} P(\mathcal{R})P_{\text{s}}(H_k \mid \mathcal{R}) \tag{3.2}$$

It is trivial to show that disruption is impossible (and that survival is certain) under "even" recombination events (De Jong 1975), so:

$$\begin{aligned} P_{\text{d}}(H_k \mid \mathcal{R}) &= 0 \quad \forall\, \mathcal{R}\ \text{even} \\ P_{\text{s}}(H_k \mid \mathcal{R}) &= 1 \quad \forall\, \mathcal{R}\ \text{even} \end{aligned}$$

Thus Eqs. 3.1 and 3.2 (the probability of disruption and survival) can be simplified to:

$$P_{\text{d}}(H_k) = \sum_{\mathcal{R}\ \text{odd}} P(\mathcal{R})P_{\text{d}}(H_k \mid \mathcal{R})$$

$$P_{\text{s}}(H_k) = \sum_{\mathcal{R}\ \text{even}} P(\mathcal{R}) + \sum_{\mathcal{R}\ \text{odd}} P(\mathcal{R})P_{\text{s}}(H_k \mid \mathcal{R})$$

This allows us to bound the probability of disruption and survival:

$$P_{\mathrm{d}}(H_k) \leq \sum_{\mathcal{R} \text{ odd}} P(\mathcal{R}) = 1 - \sum_{\mathcal{R} \text{ even}} P(\mathcal{R}) = 1 - P_{\mathrm{even}}(H_k) \quad (3.3)$$

$$P_{\mathrm{s}}(H_k) \geq \sum_{\mathcal{R} \text{ even}} P(\mathcal{R}) = P_{\mathrm{even}}(H_k) \quad (3.4)$$

Thus $P_{\mathrm{even}}(H_k)$ provides a lower bound for $P_{\mathrm{s}}(H_k)$, while $1 - P_{\mathrm{even}}(H_k)$ provides an upper bound for $P_{\mathrm{d}}(H_k)$.

For $k = 2$ the probability of drawing even recombination events under n-point recombination is:

$$P_{\mathrm{even}}(H_2, L, L_1, n) = \sum_{x=0}^{\lfloor n/2 \rfloor} \binom{n}{2x} \left(\frac{L_1}{L}\right)^{2x} \left(\frac{L - L_1}{L}\right)^{n-2x} \quad (3.5)$$

The notation $P_{\mathrm{even}}(H_2, L, L_1, n)$ means that one is computing the probability that an even number of cut-points will fall between the defining positions of H_2, under n-point recombination. The fraction L_1/L is the probability that a cut-point will fall between the two defining positions, whereas the fraction $(L - L_1)/L$ is the probability that a cut-point will fall outside the two defining positions.[3] The index $2x$ is always even, and thus the summation includes all even numbers from 0 to n. The combinatorial computes the number of ways that a particular even number of cut-points can be chosen from the n cut-points. This result was first shown by De Jong (1975), but it is valid only for second-order hyperplanes.

It is possible to generalize this to higher-order schemata by using a recurrence relation, and by noting that the sum of even numbers must be even. For $k = 3$ the probability of drawing even recombination events under n-point recombination is:

$$P_{\mathrm{even}}(H_3, L, L_1, L_2, n) =$$
$$\sum_{x=0}^{\lfloor n/2 \rfloor} \binom{n}{2x} \left(\frac{L_1}{L}\right)^{2x} \left(\frac{L - L_1}{L}\right)^{n-2x} P_{\mathrm{even}}(H_2, L_1, L_2, 2x)$$

To understand this equation, consider Fig. 3.4, in which a third-order hyperplane H_3 is depicted. The top level of the equation computes the probability that an even number of cut-points will fall between the two defining positions d_1 and d_3. The recurrence relation computes the probability that an even number of cut-points will fall between the defining positions d_1 and

[3] Note that this means that we assume that there are L possible cut-points and that they are chosen with replacement. Booker (1992) assumes $L - 1$ cut-points that are chosen without replacement; however, this makes very little difference in the computations.

d_2. Since the sum of two evens is an even, it must be the case that an even number of cut-points fall between d_2 and d_3. Thus an even number of cut-points fall between each pair of adjacent defining positions, and this equation computes the probability that an even recombination event will occur.

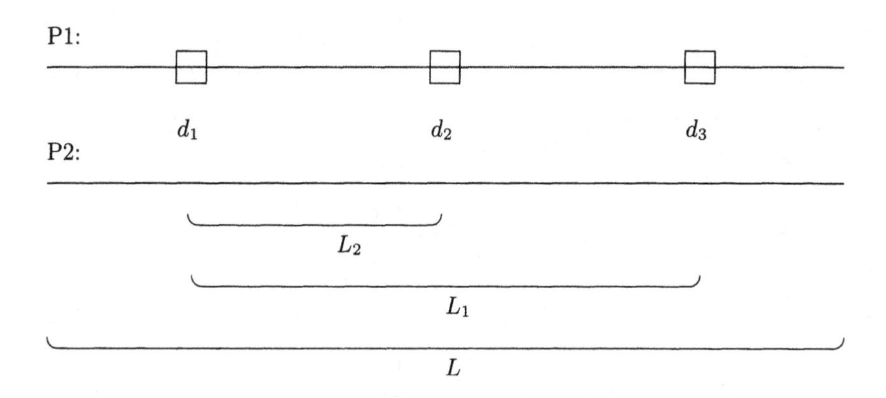

Fig. 3.4. The defining lengths of a third-order hyperplane H_3

Clearly this formulation can be extended to higher-order hyperplanes through the recurrence relation – in general the probability of drawing even recombination events under n-point recombination is:

$$P_{\text{even}}(H_k, L, L_1, ..., L_{k-1}, n) = \sum_{x=0}^{\lfloor n/2 \rfloor} \binom{n}{2x} \left(\frac{L_1}{L}\right)^{2x} \left(\frac{L - L_1}{L}\right)^{n-2x} P_{\text{even}}(H_{k-1}, L_1, ..., L_{k-1}, 2x) \tag{3.6}$$

It is important to note that the cardinality C of the alphabet has no effect on this derivation, since recombination can only swap existing alleles. As it turns out, the only time when the cardinality has an effect on the schema analysis for recombination is when population homogeneity is taken into account, as will be explained later in this chapter.

3.4 Graphing Disruption

Given Equations 3.5 and 3.6, we are now in a position to graph P_{even} for schema H_k under n-point recombination. As mentioned before, these are lower bounds on the probability of survival for H_k (see Eq. 3.4). As we know, Eqs. 3.5 and 3.6 indicate that the distances between defining positions are

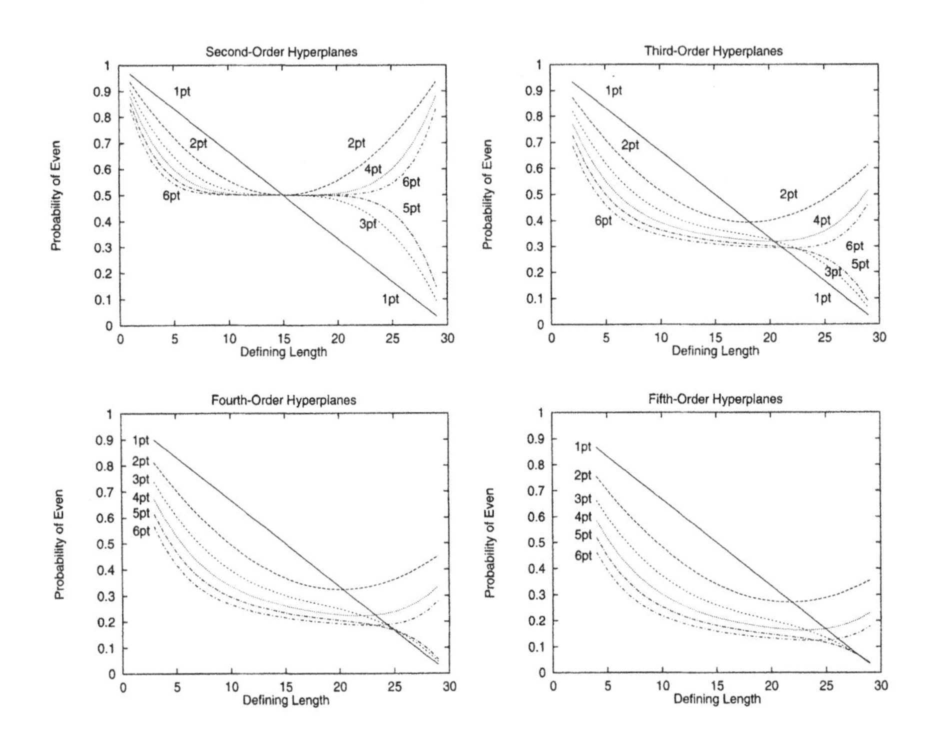

Fig. 3.5. $P_{\text{even}}(H_k)$ of H_2, H_3, H_4, and H_5 when $L = 30$, for n-point recombination

relevant when considering n-point recombination. Thus, H_k is defined by L, L_1, ... , L_{k-1}. Throughout the schema analysis in this book we assume that $L = 30$ in order to allow for comparison with Syswerda (1989) and De Jong (1975). The minimum for the defining length L_1 is $k-1$ (e.g., the closest that 3 defining positions can be is 2), and the maximum for L_1 is $L - 1 = 29$.

Figure 3.5 shows the probability that an even number of cut-points will fall between the defining positions of schema H_k, under n-point recombination (where n ranges from one to six). There are four graphs, one each for H_2, H_3, H_4, and H_5. Clearly it is simple to graph the results for H_2, since it only depends on L and L_1. However, H_3 depends also on L_2, which adds another dimension. Rather than attempt to deal with high-dimension graphs we take the alternative solution of graphing the result of averaging over all possible choices for L_2. Thus, instead of graphing $P_{\text{even}}(H_3, L, L_1, L_2, n)$ we graph:

$$P_{\text{even}}(H_3, L, L_1, n) = \frac{1}{L_1 - 1} \sum_{L_2=1}^{L_1-1} P_{\text{even}}(H_3, L, L_1, L_2, n)$$

This is also done for the graphs for H_4 and H_5, where the results are averaged over all possible choices of the defining positions. This form of averaging is used throughout this book.

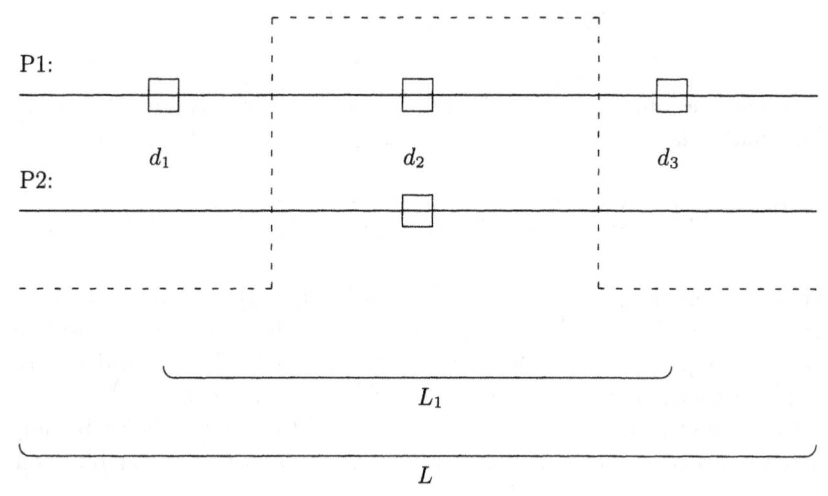

Fig. 3.6. An example of survival under an odd recombination event

Figure 3.5 illustrates a number of interesting results. For all forms of n-point recombination, P_{even} (the probability that an even number of cut-points will fall between the defining positions of H_k) is affected by both the defining lengths and the order k of the hyperplane. All forms of n-point recombination are more disruptive of higher-order hyperplanes (P_{even} gets lower as k increases). Also, all forms of n-point recombination are reasonably nondisruptive (P_{even} is reasonably high) of short schemata, but the higher n is, the greater the disruption. Finally, when the defining length is long, n-point recombination falls into two classes, depending on whether n is even or odd. When n is odd, disruption continues to increase as L_1 increases (one-point recombination acts linearly with respect to L_1, as we would expect). However, when n is even, disruption eventually decreases as L_1 increases. Thus, for the longer defining lengths, n-point recombination when n is odd is more disruptive than when n is even.

3.5 Estimates Using Population Homogeneity

As seen earlier in Eq. 3.4, $P_{\text{even}}(H_k)$ provides a lower bound for the probability of survival of a hyperplane $P_{\text{s}}(H_k)$. It is a lower bound because H_k may yet survive under an odd recombination event, if alleles are identical. This can be seen in Fig. 3.6, which shows an odd recombination event. However, the two strings have the same allele at defining position d_2, so in fact the hyperplane H_3 survives. Thus, clearly the population homogeneity plays an important role in estimating the actual probability of survival $P_{\text{s}}(H_k)$. As opposed to concentrating only on even recombination events, the computation of $P_{\text{s}}(H_k)$ examines all possible recombination events $\mathcal{R}$:

$$P_s(H_k) = \sum_{\mathcal{R}} P(\mathcal{R}) P_s(H_k \mid \mathcal{R})$$

For a second-order schemata H_2 the probability of survival under n-point recombination is:

$$P_s(H_2, L, L_1, n) = \sum_{x=0}^{n} \binom{n}{x} \left(\frac{L_1}{L}\right)^x \left(\frac{L - L_1}{L}\right)^{n-x} P_s(H_2 \mid \mathcal{R})$$

where the recombination event $\mathcal{R}$ (and hence $P_s(H_2 \mid \mathcal{R})$) is determined by x, which is the number of cut-points falling between the two defining positions of H_2. This equation can be contrasted with Eq. 3.5, which considers only *even* recombination events $\mathcal{R}$ (in which case $P_s(H_2 \mid \mathcal{R}) = 1$).

Once again the recurrence relation can be used to compute the probability of survival of a general kth-order hyperplane H_k under n-point recombination:

$$P_s(H_k, L, L_1, ..., L_{k-1}, n) = \qquad (3.7)$$
$$\sum_{x=0}^{n} \binom{n}{x} \left(\frac{L_1}{L}\right)^x \left(\frac{L - L_1}{L}\right)^{n-x} P_s(H_{k-1}, L_1, ..., L_{k-1}, x)$$

The recurrence relation ends at:

$$P_s(H_2, L, L_1, n) = \sum_{x=0}^{n} \binom{n}{x} \left(\frac{L_1}{L}\right)^x \left(\frac{L - L_1}{L}\right)^{n-x} P_s(H_k \mid \mathcal{R}) \qquad (3.8)$$

where once again $\mathcal{R}$ (and hence $P_s(H_k \mid \mathcal{R})$) is determined by how many cut-points fall between the adjacent defining positions of H_k.[4]

The goal now is to compute $P_s(H_k \mid \mathcal{R})$, which depends both on the recombination event and the population homogeneity. Consider Fig. 3.3 again. Although this is an odd recombination event, survival of H_3 occurs if the two parents have matching alleles at position d_2 or if they have matching alleles at positions d_1 and d_3. In either situation the hyperplane H_3 survives on one offspring or the other. The probability that the alleles will match depends on the population homogeneity.

For ease of presentation, let K be the set of k defining positions. Suppose that recombination results in a subset X of the k alleles from parent P1 surviving in the same offspring. In this case it is impossible for disruption to occur if: 1) the parents match on the subset X, or 2) they match on the subset $K - X$. Hence the most general form for $P_s(H_k \mid \mathcal{R})$ is:

$$P_s(H_k \mid \mathcal{R}) = P_{eq}(X) + P_{eq}(K - X) - P_{eq}(K) \qquad (3.9)$$

[4] Although this is the lowest level of the recurrence relation, we refer to $P_s(H_k \mid \mathcal{R})$ instead of $P_s(H_2 \mid \mathcal{R})$ because we care about the survival of the whole kth-order hyperplane. Also, $\mathcal{R}$ is introduced here because it is not defined until the lowest level of the recurrence relation is reached.

where $P_{eq}(X)$ represents the probability that the two parents will match on Xs alleles, while $P_{eq}(K - X)$ is the probability that the two parents will match on the remaining alleles. The third term reflects the joint probability that both parents match on all k alleles, and hence must be subtracted.[5] If an even recombination event occurs (i.e., $\mathcal{R}$ is even), then $X = K$ and $P_s(H_k \mid \mathcal{R}) = 1$. However, for an odd recombination event $P_s(H_k \mid \mathcal{R}) \leq 1$.

Deriving precise expressions for the values of $P_{eq}(X)$ at a particular point in time is difficult in general since they vary from generation to generation in complex, nonlinear, and interacting ways. We can, however, get considerable insight into the effects of shared alleles on disruption analysis by making two simplifying assumptions.

The first assumption is the independence of alleles. Let $P_{eq}(d)$ represent the probability that both parents have the same allele at a particular defining position d. Due to independence:

$$P_s(H_k \mid \mathcal{R}) = \prod_{d \in X} P_{eq}(d) + \prod_{d \in K-X} P_{eq}(d) - \prod_{d \in K} P_{eq}(d) \qquad (3.10)$$

Once again, if an even recombination event occurs, then $X = K$ and $P_s(H_k \mid \mathcal{R}) = 1$ (because the product over the null set is 1), as we would expect. However, for an odd recombination event $P_s(H_k \mid \mathcal{R}) < 1$ (unless $P_{eq}(d) = 1, \ \forall \ d \in K$).

The second assumption is that for the hyperplane H_k of interest, $P_{eq}(d)$ is identical for all the defining positions ($P_{eq}(d) = P_{eq}, \ \forall \ d \in K$).[6] Now suppose that under recombination event $\mathcal{R}$ that x of the k alleles from parent P1 survive in the same offspring. In this case it is impossible for disruption to occur if: 1) the parents match on all x defining positions, or 2) they match on all the remaining $k - x$ defining positions. In this case:

$$P_s(H_k \mid \mathcal{R}) = {P_{eq}}^x + {P_{eq}}^{k-x} - {P_{eq}}^k \qquad (3.11)$$

Under these assumptions, if an even recombination event occurs, then $x = k$ and $P_s(H_k \mid \mathcal{R}) = 1$. For an odd recombination event $P_s(H_k \mid \mathcal{R}) < 1$ unless $P_{eq} = 1$, as we would expect.

If $P_{eq} = 0.0$, then $P_s(H_k \mid \mathcal{R}) = 0$ for odd events $\mathcal{R}$, so the previous graphs of P_{even} (see Fig. 3.5) are equivalent to graphing $P_s(H_k)$ when $P_{eq} = 0.0$. We can now see how P_{eq} affects the probability of survival by examining Figs. 3.7 and 3.8, which graph the probability of survival $P_s(H_k)$ when $P_{eq} = 0.5$ and $P_{eq} = 0.75$. For a randomly initialized population any allele has probability $1/C$ of being the same as any other allele, so the minimum P_{eq} is simply $1/C$. A value close to 0.5 is a reasonable estimate for a randomly initialized

[5] If A and B are events, then $P(A \vee B) = P(A) + P(B) - P(A \wedge B)$.

[6] The two assumptions of independence and identicalness are so useful for illustrating the results that they will be used (often implicitly) throughout the static analyses performed in this book. Syswerda (1989) also implicitly makes these two assumptions and denotes P_{eq} as P_{ok}, but only considers $P_{eq} = P_{ok} = 0.5$.

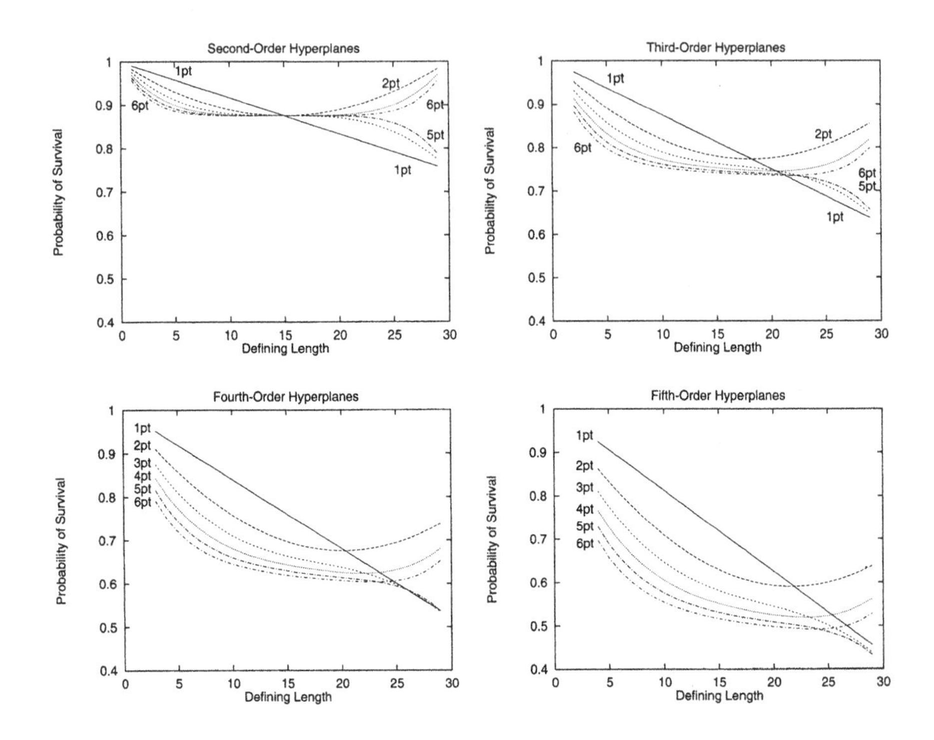

Fig. 3.7. $P_s(H_k)$ of H_2, H_3, H_4, and H_5 when $L = 30$ and $P_{eq} = 0.5$, for n-point recombination

population consisting of binary strings ($C = 2$). A value close to 0 is a reasonable estimate for a randomly initialized population consisting of strings from a high-cardinality alphabet. Since the population of a traditional EA tends to become more homogeneous with time, values of $P_{eq} > 1/C$ represent a population that is converging.[7]

A comparison of Fig. 3.7 and Fig. 3.8 with Fig. 3.5 indicates that as P_{eq} increases, so does the probability of survival. This is reasonable, since odd recombination events become less likely to cause disruption. Despite these global changes to the curves, however, the basic relationships between the curves remain the same. For example, for the longer defining lengths, n-point recombination when n is odd is still more disruptive than when n is even.

It is important to emphasize that the two assumptions of independence and identicalness have been made in order to gain insight into the disruptive effect that recombination has on hyperplanes H_k. These assumptions are valid when the population has been randomly initialized, but generally become less valid as the population evolves. Despite this, the insights gained are often

[7] This is the only effect that the cardinality C has on the analysis, namely, it determines what values of P_{eq} are of interest.

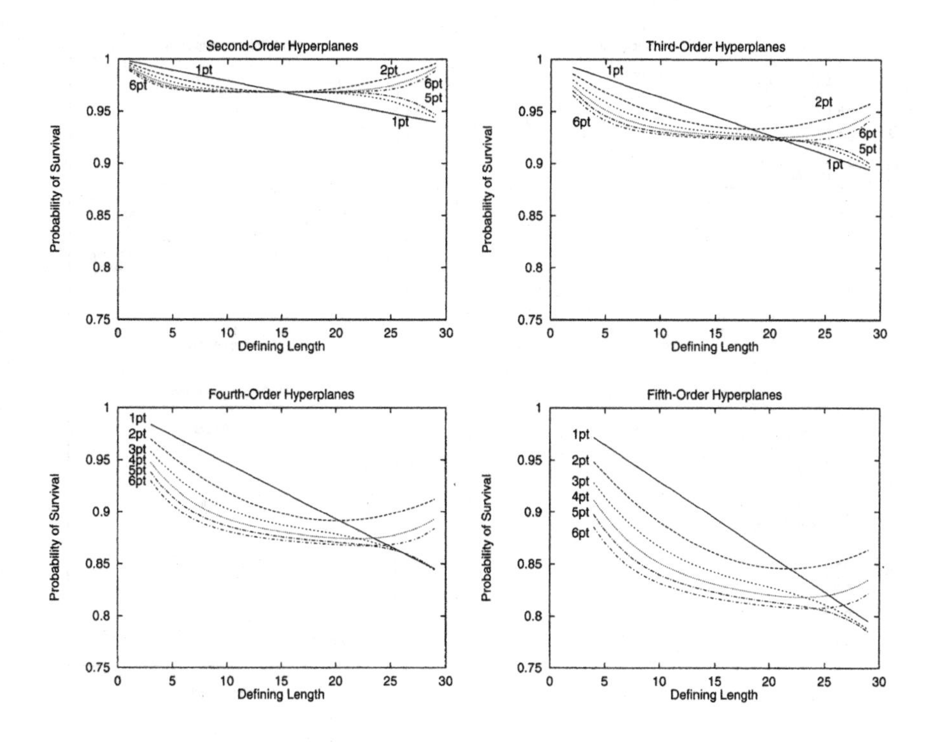

Fig. 3.8. $P_s(H_k)$ of H_2, H_3, H_4, and H_5 when $L = 30$ and $P_{eq} = 0.75$, for n-point recombination

quite valuable, as will be demonstrated in the dynamic analyses of this book (starting at Chap. 10), where these two assumptions are dropped.

3.6 Survival Theory for P_0 Uniform Recombination

In uniform recombination alleles are exchanged between two parents with probability P_0. In this context an even recombination event occurs only if all k defining positions of H_k are exchanged or if all k defining positions are not exchanged. This can also be described in terms of bit masks. An even event corresponds to a bit mask of k 1s or k 0s. Thus the probability of an even recombination event is simply:

$$P_{\text{even}}(H_k, P_0) \;=\; {P_0}^k \;+\; (1 - P_0)^k \tag{3.12}$$

As before, we can obtain an exact formulation of $P_s(H_k, P_0)$ if we include the nondisruptive odd recombination events. For uniform recombination this corresponds to those bit masks which are not either all 0s or all 1s on the hyperplane defining positions, but are nondisruptive because the parents share common alleles on those particular positions. Once again denote K as the set

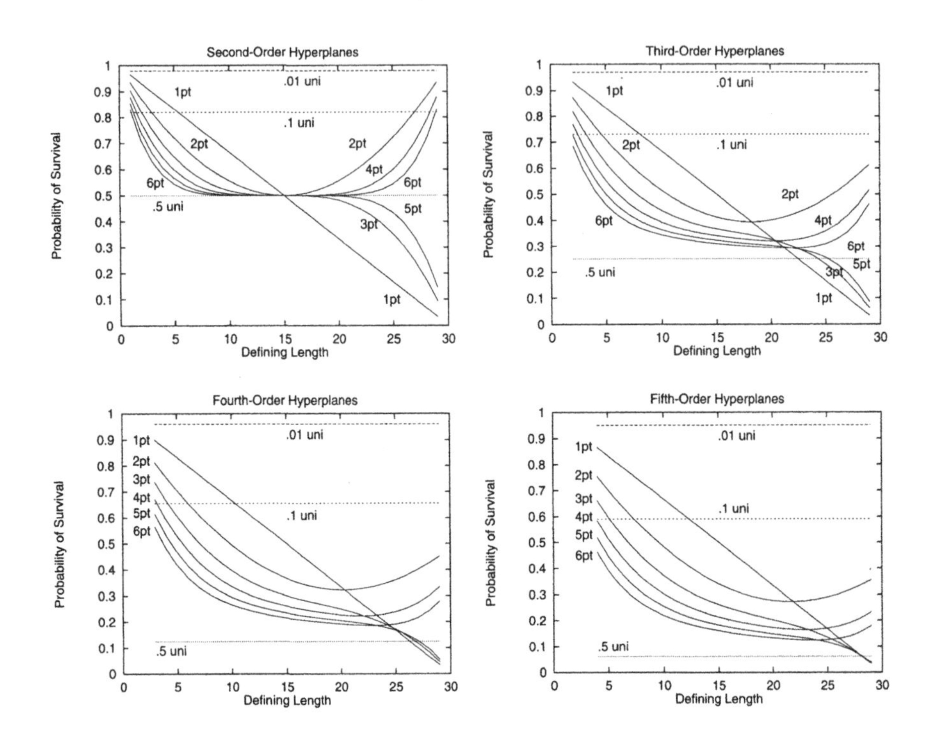

Fig. 3.9. $P_s(H_k)$ of H_2, H_3, H_4, and H_5 when $L = 30$ and $P_{eq} = 0.0$, for P_0 uniform recombination

of k defining positions. Let $|K|$ be the cardinality of the set K, and $PS(K)$ be the power set of K. Then the probability of survival for H_k under P_0 uniform recombination is:

$$P_s(H_k, P_0) = \sum_{X \in PS(K)} (P_0)^{|X|}(1 - P_0)^{|K-X|} P_s(H_k \mid \mathcal{R})$$

where $\mathcal{R}$ is determined by the set X of alleles that survived in the same offspring. Once again, the goal is to compute $P_s(H_k \mid \mathcal{R})$, which is a function of population homogeneity. In this case no disruption will occur if: 1) the parents match on the subset X, or 2) they match on the subset $K-X$. Hence, just as with n-point recombination, the most general form for $P_s(H_k \mid \mathcal{R})$ is:

$$P_s(H_k \mid \mathcal{R}) = P_{eq}(X) + P_{eq}(K - X) - P_{eq}(K)$$

where $P_{eq}(X)$ represents the probability that the two parents will match on Xs alleles, while $P_{eq}(K - X)$ is the probability that the two parents will match on the remaining alleles. As before, the third term reflects the joint probability that both parents match on all k alleles, and hence must be subtracted.

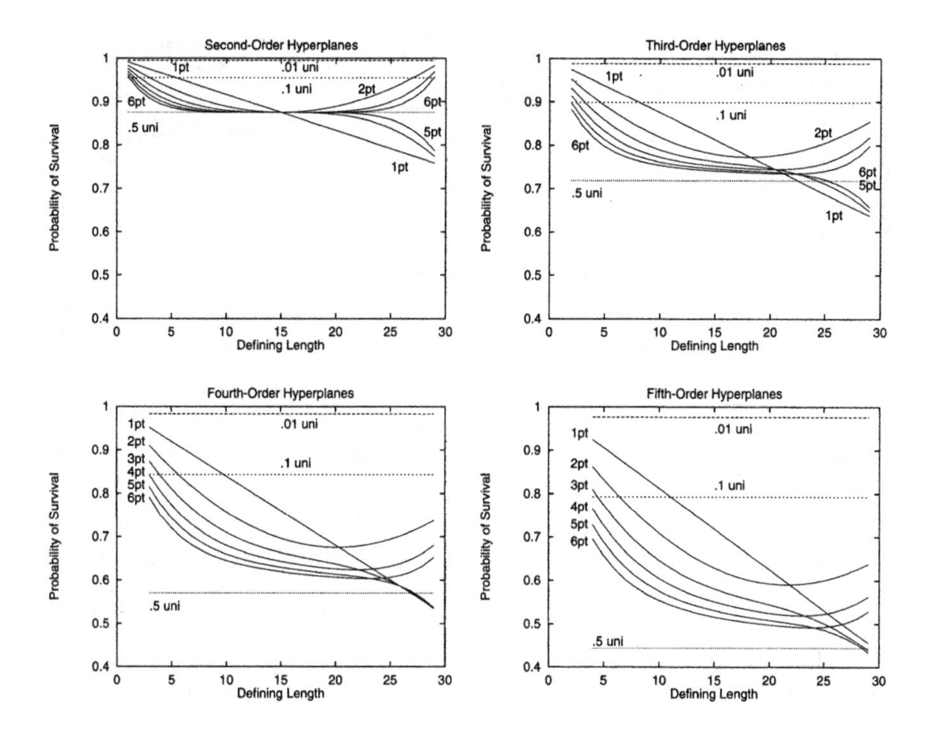

Fig. 3.10. $P_s(H_k)$ of H_2, H_3, H_4, and H_5 when $L = 30$ and $P_{eq} = 0.5$, for P_0 uniform recombination

As with n-point recombination, assumptions concerning independence and identicalness are useful. Let $P_{eq}(d)$ represent the probability that both parents have the same allele at a particular defining position d. Then independence implies:

$$P_s(H_k \mid \mathcal{R}) = \prod_{d \in X} P_{eq}(d) + \prod_{d \in K-X} P_{eq}(d) - \prod_{d \in K} P_{eq}(d)$$

A further assumption that $P_{eq}(d)$ is identical for all defining positions ($P_{eq}(d) = P_{eq}, \ \forall \ d \in K$) implies:

$$P_s(H_k \mid \mathcal{R}) = {P_{eq}}^x + {P_{eq}}^{k-x} - {P_{eq}}^k$$

Given the two assumptions of independence and identicalness, the probability of survival of H_k under P_0 uniform recombination is simply:

$$P_s(H_k, P_0) = \tag{3.13}$$
$$\sum_{x=0}^{k} \binom{k}{x} {P_0}^x (1 - P_0)^{k-x} ({P_{eq}}^x + {P_{eq}}^{k-x} - {P_{eq}}^k)$$

Figures 3.9, 3.10, and 3.11 graph the probability of survival $P_s(H_k)$ when $P_{eq} = 0.0$, $P_{eq} = 0.5$, and $P_{eq} = 0.75$. Because the disruptive effects of P_0

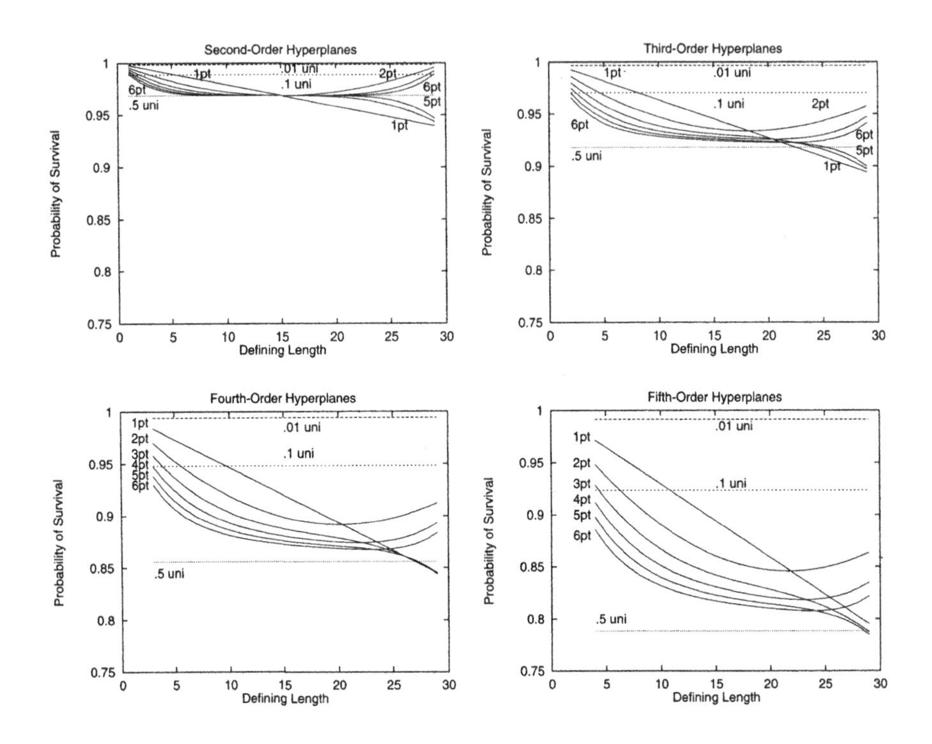

Fig. 3.11. $P_s(H_k)$ of H_2, H_3, H_4, and H_5 when $L = 30$ and $P_{eq} = 0.75$, for P_0 uniform recombination

uniform recombination are symmetric around 0.5 (e.g., 0.2 uniform recombination is the same as 0.8 uniform recombination), it suffices to examine P_0 in the range of 0.0 to 0.5. The previous results from n-point recombination are also included for the sake of comparison.

The graphs indicate that, as expected, P_0 uniform recombination is only affected by the order k of the hyperplanes – it is not affected by defining lengths (unlike n-point recombination). Also, 0.5 uniform recombination is the most disruptive setting for P_0 uniform recombination, and in fact it is (in general) more disruptive than n-point recombination (especially for higher-order hyperplanes). Furthermore, the amount of disruption caused by P_0 uniform recombination can be lowered by simply reducing P_0 – when $P_0 = 0.0$ there is no disruption at all. These relative results do not appear to change as the population homogeneity P_{eq} changes.

Syswerda (1989) provides a similar graph for $P_s(H_5)$ when $P_{eq} = 0.5$, for one-point, two-point, and 0.5 uniform recombination. Despite the fact that Syswerda uses a different derivation, the curves in his graph agree with those here, providing independent confirmation of the results presented thus far in this book.

3.6.1 A Special Case: $P_{eq} = 0.0$

As an interesting check on Eq. 3.13, consider the case where there is maximum population diversity ($P_{eq} = 0.0$). In that case (${P_{eq}}^x + {P_{eq}}^{k-x} - {P_{eq}}^k$) is nonzero only when $x = 0$ and $x = k$, and:

$$P_s(H_k, P_0) = \qquad (3.14)$$
$$ {P_0}^0 \, (1 - P_0)^k \; + \; {P_0}^k \, (1 - P_0)^0 \; = \; (1 - P_0)^k \; + \; {P_0}^k$$

Thus, as would be expected, survival can only occur if all k alleles are exchanged, or if all k are not. Clearly, this is equivalent to $P_{\text{even}}(H_k, P_0)$ in Eq. 3.12.

3.6.2 A Special Case: $P_0 = 0.0$

As pointed out above, one interesting feature of P_0 uniform recombination is that it can be turned off simply by setting P_0 to 0.0. If $P_0 = 0.0$ the only term in Eq. 3.13 that is not zero is when $x = 0$. In that case:

$$P_s(H_k, P_0 = 0.0) \; = \; {P_{eq}}^0 \; + \; {P_{eq}}^k \; - \; {P_{eq}}^k \; = \; 1.0 \qquad (3.15)$$

This is as expected, since when uniform recombination is turned off the hyperplane H_k must survive.

3.6.3 A Special Case: $P_0 = 0.5$

Another simplification to Eq. 3.13 can be made if $P_0 = 0.5$. In that case each recombination event occurs with probability $1/2^k$ and:

$$P_s(H_k, P_0 = 0.5) \; = \; \frac{1}{2^k} \sum_{x=0}^{k} \binom{k}{x} ({P_{eq}}^x \; + \; {P_{eq}}^{k-x} \; - \; {P_{eq}}^k)$$

Expanding yields:

$$P_s(H_k, P_0 = 0.5) \; =$$
$$\frac{1}{2^k} \left[\sum_{x=0}^{k} \binom{k}{x} {P_{eq}}^x \; + \; \sum_{x=0}^{k} \binom{k}{x} {P_{eq}}^{k-x} \; - \; {P_{eq}}^k \sum_{x=0}^{k} \binom{k}{x} \right]$$

This can be manipulated into a form amenable to the binomial expansion:

$$P_s(H_k, P_0 = 0.5) \; = \; \frac{1}{2^k} \times$$
$$\left[\sum_{x=0}^{k} \binom{k}{x} {P_{eq}}^x 1^{k-x} + \sum_{x=0}^{k} \binom{k}{x} 1^x {P_{eq}}^{k-x} - {P_{eq}}^k \sum_{x=0}^{k} \binom{k}{x} 1^x 1^{k-x} \right]$$

Now applying the binomial expansion yields:

$$P_s(H_k, P_0 = 0.5) = \frac{1}{2^k}\left[(1+P_{eq})^k + (1+P_{eq})^k - 2^k {P_{eq}}^k\right]$$

or:

$$P_s(H_k, P_0 = 0.5) = \frac{2(1+P_{eq})^k}{2^k} - {P_{eq}}^k \qquad (3.16)$$

Since $P_0 = 0.5$ is the most commonly used setting for P_0 uniform recombination, Eq. 3.16 is very useful, providing a simple equation for computing the probability of survival of a schema H_k.

3.7 Expected Number of Offspring in H_k

Thus far this chapter has provided a thorough analysis of the probability that a hyperplane will survive recombination, for n-point and P_0 uniform recombination. However, as stated at the beginning of this chapter, a comparison of recombination with mutation will require the computation of the expected number of offspring in a hyperplane. Thus, the goal now is to convert the probability analysis in this chapter into an analysis yielding the expected number of offspring that will reside in H_k after recombination has occurred.

In the analyses by both Holland (1975) and De Jong (1975) the probability of survival (disruption) was based on an implicit assumption that $P_{eq} = 0.0$. In such cases a disruption must occur if an odd number of cut-points fall between any adjacent pair of defining positions of the hyperplane H_k. Thus only one parent can be an instance of H_k, and it is impossible for both parents to be instances of H_k. Similarly, at most one offspring can be an instance of H_k, and it is impossible for both offspring to be in H_k. Due to this restriction, the expected number of offspring in H_k is precisely given by $P_s(H_k)$. However, once P_{eq} can differ from 0, it now becomes possible for both parents to be instances of H_k. If that happens then both offspring will be instances of H_k as well. This complicates the computation of the expected number of offspring in H_k somewhat, as we explore in this section.

Consider A to be a random variable that describes the number of parents (of the two parents considered for recombination) that are instances of H_k. For survival analysis it is always assumed that at least one parent is an instance of H_k, so A can take on values 1 and 2. Let B be a random variable describing the number of offspring that are instances of H_k. B can take on values 0, 1, and 2. We can write:

$$P_s(H_k) = P(B = 1 \vee 2 \mid A = 1 \vee 2)$$
$$P_d(H_k) = P(B = 0 \mid A = 1 \vee 2)$$

The expression for $P_s(H_k)$ can be expanded:

$$P_s(H_k) = \\ P(B=1 \wedge A=1) + P(B=1 \wedge A=2) + \\ P(B=2 \wedge A=1) + P(B=2 \wedge A=2)$$

$$P_s(H_k) = \\ P(B=1 \mid A=1)P(A=1) + P(B=1 \mid A=2)P(A=2) + \\ P(B=2 \mid A=1)P(A=1) + P(B=2 \mid A=2)P(A=2)$$

The second and third terms are zero, because if two parents are instances of H_k, it is impossible for only one offspring to be an instance of H_k (and vice versa). Also, if both parents are in H_k, both offspring must be in H_k. Thus we get:

$$P_s(H_k) = P(B=1 \mid A=1)P(A=1) + P(A=2)$$

Now the expected number of offspring in H_k can be computed as follows:

$$E[B] = \sum_{b \in \{0,1,2\}} b \times P(B=b) = P(B=1) + 2P(B=2)$$

For the sake of clarity, denote $E[B]$ to be $E_s[B_k]$. $E_s[B_k]$ will refer to the expected number of offspring that will be in H_k. The 's' subscript is a reminder that the situation is one in which the survival of an individual in H_k is at stake.

$$E_s[B_k] = \\ P(B=1 \mid A=1)P(A=1) + P(B=1 \mid A=2)P(A=2) + \\ 2[P(B=2 \mid A=1)P(A=1) + P(B=2 \mid A=2)P(A=2)]$$

Again, two of the terms disappear:

$$E_s[B_k] = P(B=1 \mid A=1)P(A=1) + 2P(A=2)$$

Thus:

$$E_s[B_k] = P_s(H_k) + P(A=2)$$

Now the probability that two parents are in H_k is controlled by P_{eq} (assuming independence and identicalness, as stated earlier). Clearly $P(A=2) = {P_{eq}}^k$, so:

$$E_s[B_k] = P_s(H_k) + {P_{eq}}^k \tag{3.17}$$

So, we can see that the computation of the expected number of offspring that will reside in H_k after recombination is easily computed from $P_s(H_k)$ and the population homogeneity P_{eq}. Notice that if $P_{eq} = 0.0$, $E_s[B_k] = P_s(H_k)$, which is what we expected. Notice also that as P_{eq} approaches one both $P_s(H_k)$ and ${P_{eq}}^k$ approach one, so $E_s[B_k]$ approaches two as we would expect.

For a particular hyperplane H_k and population homogeneity P_{eq}, the previous graphs would only need to be offset by ${P_{eq}}^k$ to produce graphs for $E_s[B_k]$. Since the offset would be the same for each recombination operator within each graph, the relationships between the different recombination operators would not change.

3.8 Summary

This chapter first computes the probability $P_s(H_k)$ that a kth-order hyperplane H_k will survive recombination, given that one parent is a member of H_k and that the other parent is arbitrary. Both n-point and P_0 uniform recombination are analyzed. The cardinality of the alphabet and the population homogeneity are taken into account in a natural fashion. The results indicate that the disruptive aspect of n-point recombination is affected by both the defining lengths and the order k of the hyperplane, while P_0 uniform recombination is only affected by the order. All forms of recombination are more disruptive of higher-order hyperplanes and become less disruptive when the population converges.

All forms of n-point recombination are quite nondisruptive of short schemata, but the higher n is, the greater the disruption. For long schemata, n-point recombination when n is odd is more disruptive than when n is even.

The disruption caused by P_0 uniform recombination varies with P_0. The most disruptive setting for P_0 uniform recombination is when $P_0 = 0.5$. In general this is more disruptive than n-point recombination (especially for higher-order hyperplanes). Finally, the amount of disruption caused by P_0 uniform recombination can be lowered by simply reducing P_0 – when $P_0 = 0.0$ there is no disruption at all.

This chapter concludes by computing the expected number of offspring $E_s[B_k]$ that reside in H_k after recombination, and shows that this is a simple function of $P_s(H_k)$, the population homogeneity P_{eq}, and the order of the hyperplane k. The computation of $E_s[B_k]$ is important because it will allow for a fair comparison with mutation (in Chaps. 6–8).

As pointed out in the beginning of this chapter, hyperplanes need not only survive – they also may be constructed from other hyperplanes. This effect is analyzed in the next chapter.

4. A Construction Schema Theory for Recombination

4.1 Introduction

Chapter 3 computed the probability that a hyperplane H_k will survive recombination, given that one parent is in H_k and that the other parent is arbitrary. However, as Syswerda (1989) pointed out, recombination can also be considered to have a more positive role – that of *construction.* Construction refers to having recombination create an instance of a hyperplane H_k from both parents. As we will see, construction is a more general concept than simple survival.

Assume that the following random experiment is being performed. One is given two parents, and one parent is in the hyperplane H_m, while the other parent is in the hyperplane H_n. Figure 4.1 provides a pictorial example. The two parents are labeled P1 and P2. P1 is a member of a particular second-order hyperplane H_2, described by the alleles at defining positions d_1 and d_2. P2 is a member of another second-order hyperplane H_2, described by the alleles at defining positions d_3 and d_4. The goal is to compute the probability that recombination will produce an offspring that is in the fourth-order hyperplane H_4 which has those four alleles at the defining positions d_1, d_2, d_3, and d_4. Thus if P1 is a member of AA## and P2 is a member of ##AA, the goal is to compute the probability that AAAA will be constructed.

In the remaining discussion we will consider the creation of a kth-order hyperplane H_k from two hyperplanes of order m (H_m) and order n (H_n). We will restrict the situation such that the two lower-order hyperplanes H_m and H_n are nonoverlapping, and $k = m + n$. Each lower-order hyperplane is represented by a different parent.

These situations will be described by random variable $\mathcal{S}$. For a kth-order hyperplane H_k there are 2^k possible situation events: $0 \leq \mathcal{S} \leq 2^k - 1$. Each situation event $\mathcal{S}$ can be represented by a bit mask of length k. The binary representation of $\mathcal{S}$ indicates which parent has each of the k alleles. There will be m 1s and n 0s in the binary representation of $\mathcal{S}$, indicating H_m and H_n. For example, the situation described in Fig. 4.1 can be described with the binary string 1100 (or 0011), to indicate that one parent has the first two alleles, while the second parent has the second two alleles of the fourth-order hyperplane.

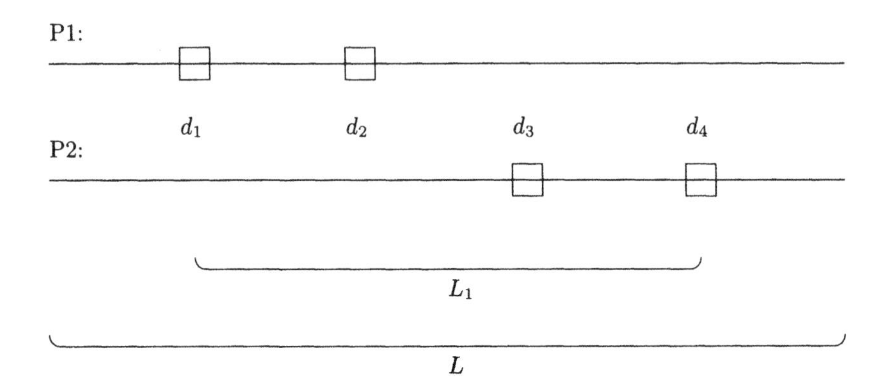

Fig. 4.1. The setup for the random experiment to be performed. P1 is a member of a second-order hyperplane, while P2 is a member of another second-order hyperplane. The goal is to construct the fourth-order hyperplane with recombination.

We denote the probability that the kth-order hyperplane will be recombined from the two hyperplanes H_m and H_n as $P_c(H_k \mid \mathcal{S}) \equiv P_c(H_k \mid H_m \wedge H_n)$, where $\mathcal{S}$ uniquely defines the two hyperplanes H_m and H_n.

4.2 Construction Theory for n-point Recombination

Equations 3.7 and 3.8 of Chap. 3 computed the probability that a hyperplane H_k will survive n-point recombination by using a recurrence relation on the order of the hyperplane (until the bottom level of a second-order hyperplane is reached). A similar formulation can be used to compute the probability of construction of H_k under n-point recombination:

$$P_c(H_k, L, L_1, ..., L_{k-1}, n \mid \mathcal{S}) = \sum_{x=0}^{n} \binom{n}{x} \left(\frac{L_1}{L}\right)^x \left(\frac{L-L_1}{L}\right)^{n-x} P_c(H_{k-1}, L_1, ..., L_{k-1}, x \mid \mathcal{S}) \quad (4.1)$$

where, for $k = 2$ the probability of construction under n-point recombination is:

$$P_c(H_2, L, L_1, n \mid \mathcal{S}) = \sum_{x=0}^{n} \binom{n}{x} \left(\frac{L_1}{L}\right)^x \left(\frac{L-L_1}{L}\right)^{n-x} P_c(H_k \mid \mathcal{S} \wedge \mathcal{R}) \quad (4.2)$$

The effect of the recurrence relation is to consider all possible ways of placing the n cut-points between the k defining points of H_k. The recombination event $\mathcal{R}$ is determined by how many cut-points fall between the adjacent defining positions of H_k. Given the situation $\mathcal{S}$ and the recombination event $\mathcal{R}$, the goal is to compute $P_c(H_k \mid \mathcal{S} \wedge \mathcal{R})$.[1] As with the analogous discussion (concerning the probability of survival) in Chap. 3, this computation clearly depends on population homogeneity, since as the population becomes more homogeneous, construction of the hyperplane H_k becomes more likely.

In point of fact, the line of reasoning is essentially identical to that followed in Chap. 3. Recall that construction will occur if both lower-order hyperplanes survive in the same individual. Let K be the set of k defining positions in H_k. Suppose that recombination results in a subset X of the k alleles surviving in the same individual. In this case construction will occur if: 1) the parents match on the subset X, or 2) they match on the subset $K - X$. Hence the most general form for $P_c(H_k \mid \mathcal{S} \wedge \mathcal{R})$ is:

$$P_c(H_k \mid \mathcal{S} \wedge \mathcal{R}) \;=\; P_{eq}(X) \;+\; P_{eq}(K - X) \;-\; P_{eq}(K)$$

where $P_{eq}(X)$ represents the probability that the two parents will match on Xs alleles, while $P_{eq}(K - X)$ is the probability that the two parents will match on the remaining alleles. The third term reflects the joint probability that both parents match on all k alleles, and hence must be subtracted.

As with Chap. 3, since the computation of $P_{eq}(X)$ can often be difficult, considerable insight can be gained by assuming independence between alleles. In this case:

$$P_c(H_k \mid \mathcal{S} \wedge \mathcal{R}) \;=\; \prod_{d \in X} P_{eq}(d) \;+\; \prod_{d \in K-X} P_{eq}(d) \;-\; \prod_{d \in K} P_{eq}(d)$$

As usual, a further simplification occurs if one assumes identicalness: $P_{eq}(d) = P_{eq},\ \forall\ d \in K$. If recombination results in x of the k defining positions surviving in the same individual (i.e., x is a subset of the $k = m + n$ defining positions), then construction will occur if: 1) the parents match on all of the x positions, or 2) if they match on all $k - x$ positions. In this case $P_c(H_k \mid \mathcal{S} \wedge \mathcal{R})$ is more simply expressed as:

$$P_c(H_k \mid \mathcal{S} \wedge \mathcal{R}) \;=\; P_{eq}{}^{x} \;+\; P_{eq}{}^{k-x} \;-\; P_{eq}{}^{k} \qquad (4.3)$$

As an example, consider Fig. 4.2. In this figure we represent the recombination of two second-order hyperplanes. For the given recombination event depicted (by the dashed line), three of the alleles at the four defining positions will survive on individual P1. However, construction of the fourth-order hyperplane will still occur if the two parents match at defining position d_4 or if they match at the remainder of the defining positions. The probability of this is $P_{eq}{}^{1} + P_{eq}{}^{3} - P_{eq}{}^{4}$.

[1] Although this is the lowest level of the recurrence relation, we refer to $P_c(H_k \mid \mathcal{S} \wedge \mathcal{R})$ instead of $P_c(H_2 \mid \mathcal{S} \wedge \mathcal{R})$ because we care about the construction of the whole kth-order hyperplane. Also, $\mathcal{R}$ is introduced here because it is not defined until the lowest level of the recurrence relation is reached.

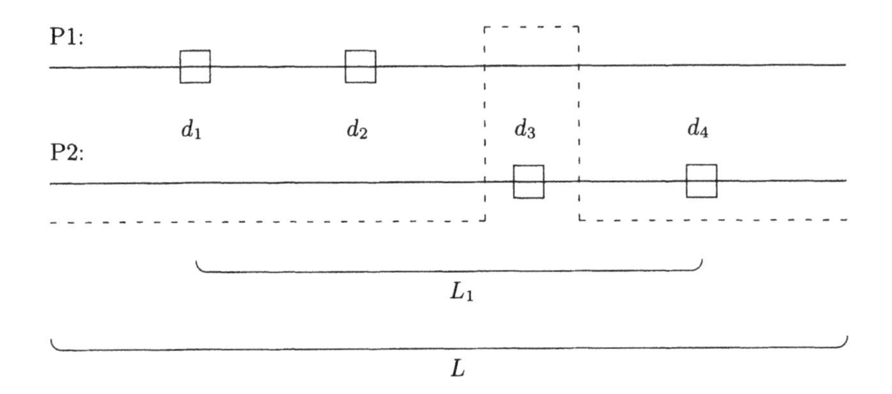

Fig. 4.2. Even if recombination does not place all four alleles on the same parent, construction of the fourth-order hyperplane will still occur if the parents match alleles at defining position d_4.

4.3 Graphing Construction

As mentioned above, we considered the creation of a kth-order hyperplane H_k from the following "situation": two hyperplanes H_m and H_n of order m and n are constructed to form H_k, where the two lower-order hyperplanes are nonoverlapping and $k = m + n$. Each lower-order hyperplane is represented by a different parent.

There are 2^k situations, since each allele at the defining positions of H_k is represented by one parent or the other. Situations are represented by $0 \le \mathcal{S} \le 2^k - 1$, in which the binary representation of $\mathcal{S}$ represents which parent has which defining position. There will be m 1s and n 0s in the binary representation of $\mathcal{S}$, indicating H_m and H_n.

It is important to note that if $\mathcal{S} = 0$ then this represents the special situation in which $H_n = H_k$. In this situation there is no construction; it in fact is merely survival. If $\mathcal{S} = 2^k - 1$ then $H_m = H_k$ and again there is no construction, but merely survival. Thus $\mathcal{S} = 0$ and $\mathcal{S} = 2^k - 1$ represent survival situations, and:

$$P_c(H_k \mid \mathcal{S} = 0) \;=\; P_c(H_k \mid \mathcal{S} = 2^k - 1) \;=\; P_s(H_k)$$

All the other situations $(0 < \mathcal{S} < 2^k - 1)$ represent true constructions, in which part of H_k is represented by one parent, while the remainder is represented by the other parent.

In order to graph the probability of construction, we consider only the $2^k - 2$ construction situations (leaving out the probability of simply surviving). Each of the $2^k - 2$ situations is considered to be equally likely, thus an average probability of construction is given by:

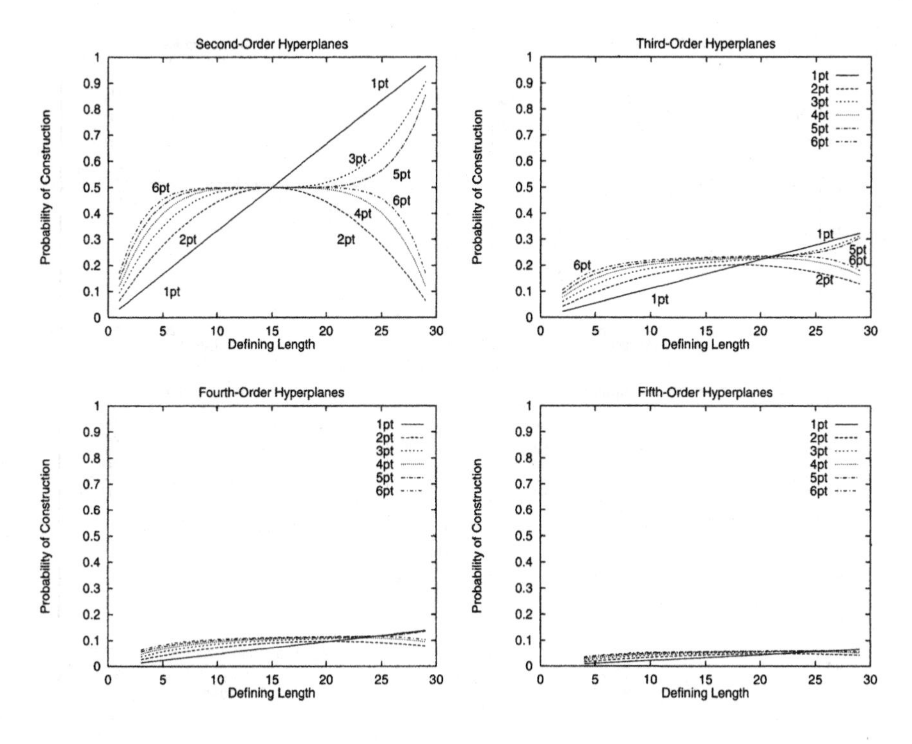

Fig. 4.3. $P_c(H_k)$ of H_2, H_3, H_4, and H_5 when $L = 30$ and $P_{eq} = 0.0$, for n-point recombination

$$P_c(H_k) = \frac{1}{2^k - 2} \sum_{\mathcal{S}=1}^{2^k-2} P_c(H_k \mid \mathcal{S}) \tag{4.4}$$

Figures 4.3, 4.4, and 4.5 graph the probability of construction $P_c(H_k)$ when $P_{eq} = 0.0$, $P_{eq} = 0.5$, and $P_{eq} = 0.75$. The results indicate that the probability of construction is affected by both the defining lengths and the order k of the hyperplane. All forms of n-point recombination are reasonably nonconstructive of short schemata, but the higher n is, the greater the construction. For long schemata, n-point recombination when n is odd is more constructive than when n is even. As P_{eq} increases, so does the probability of construction. One can see, however, that despite these global changes to the curves, the basic relationships between the curves remain the same.

Another interesting observation is that the graphs for construction are qualitatively the opposite of the graphs for survival given in Chap. 3. That is, if a hyperplane is more likely to survive under a particular n-point recombination operator then it is less likely to be constructed with that recombination operator. This would appear to make sense, since a more disruptive operator would appear to have a higher likelihood of construction. In the next section

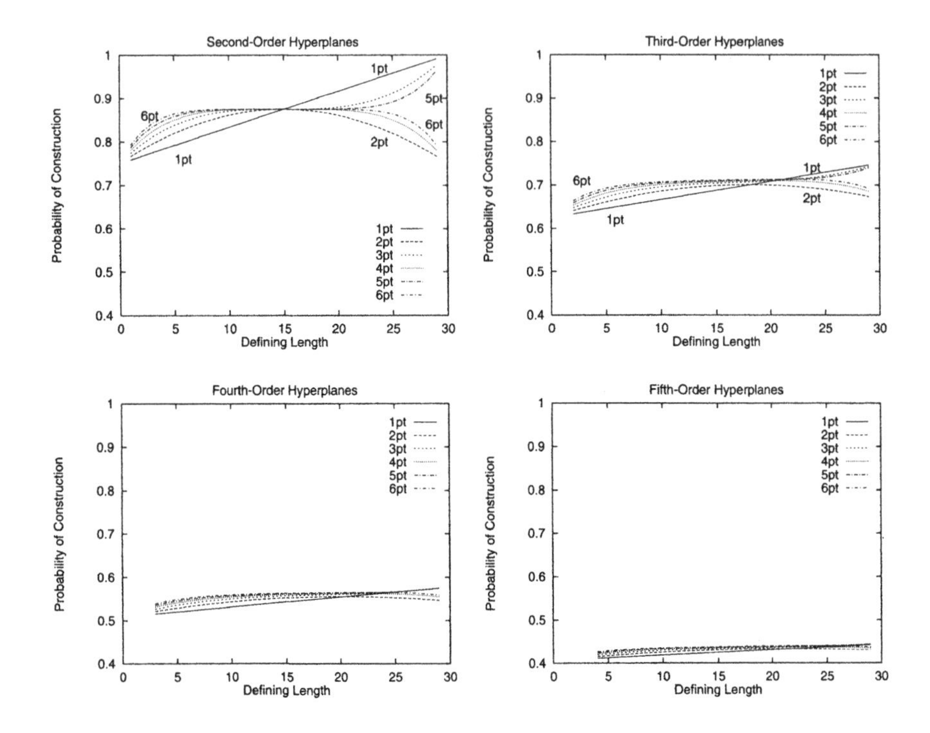

Fig. 4.4. $P_c(H_k)$ of H_2, H_3, H_4, and H_5 when $L = 30$ and $P_{eq} = 0.5$, for n-point recombination

we investigate whether this observation also holds for P_0 uniform recombination.

4.4 Construction Theory for P_0 Uniform Recombination

Chapter 3 showed that the probability of survival of a kth-order hyperplane H_k under P_0 uniform recombination is:

$$P_s(H_k, P_0) = \sum_{X \in PS(K)} (P_0)^{|X|}(1 - P_0)^{|K-X|}(P_{eq}(X) + P_{eq}(K - X) - P_{eq}(K))$$

Note that this equation can be divided into three parts. The first part can be considered to express the probability that a hyperplane H_k will survive in the original string:

$$P_{s,orig}(H_k, P_0) = \sum_{X \in PS(K)} (P_0)^{|X|}(1 - P_0)^{|K-X|} P_{eq}(X)$$

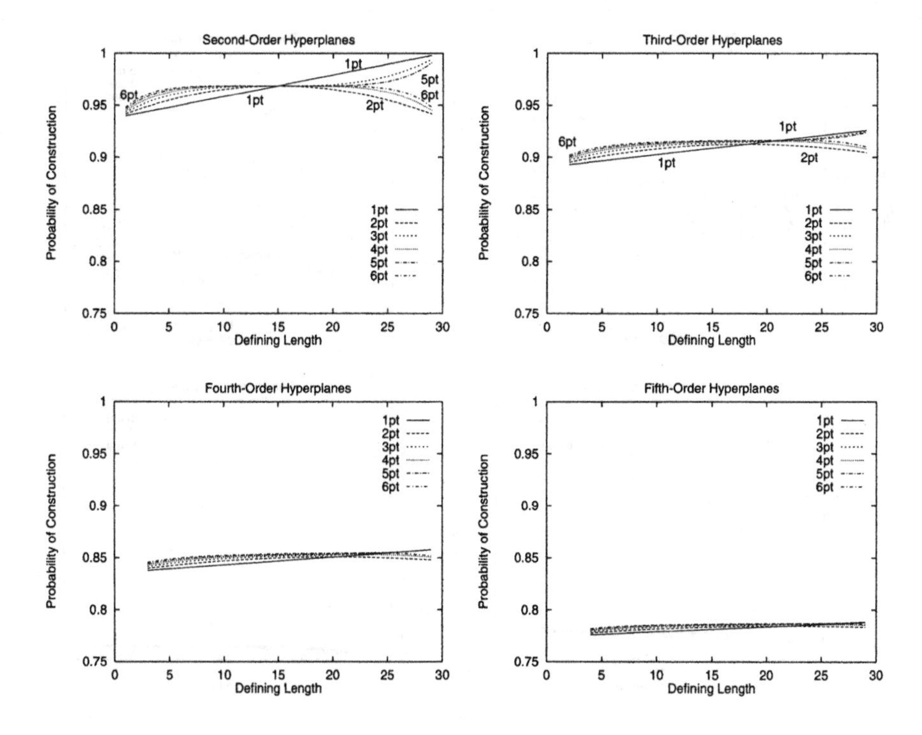

Fig. 4.5. $P_c(H_k)$ of H_2, H_3, H_4, and H_5 when $L = 30$ and $P_{eq} = 0.75$, for n-point recombination

The second part expresses the probability that a hyperplane H_k will survive in the other string:

$$P_{\text{s,other}}(H_k, P_0) \;=\; \sum_{X \in PS(K)} (P_0)^{|X|}(1 - P_0)^{|K-X|} P_{\text{eq}}(K - X)$$

The final part expresses the probability that a hyperplane H_k will exist in both strings:

$$P_{\text{s,both}}(H_k, P_0) \;=\; \sum_{X \in PS(K)} (P_0)^{|X|}(1 - P_0)^{|K-X|} P_{\text{eq}}(K) \;=\; P_{\text{eq}}(K)$$

Then:

$$P_{\text{s}}(H_k, P_0) \;=\; P_{\text{s,orig}}(H_k, P_0) \;+\; P_{\text{s,other}}(H_k, P_0) \;-\; P_{\text{s,both}}(H_k, P_0)$$

What is nice about this formulation is that it allows us to easily compute the probability of construction of H_k under P_0 uniform recombination. Assuming the construction of two nonoverlapping hyperplanes of order n and m into a hyperplane of order k, the probability of construction is:

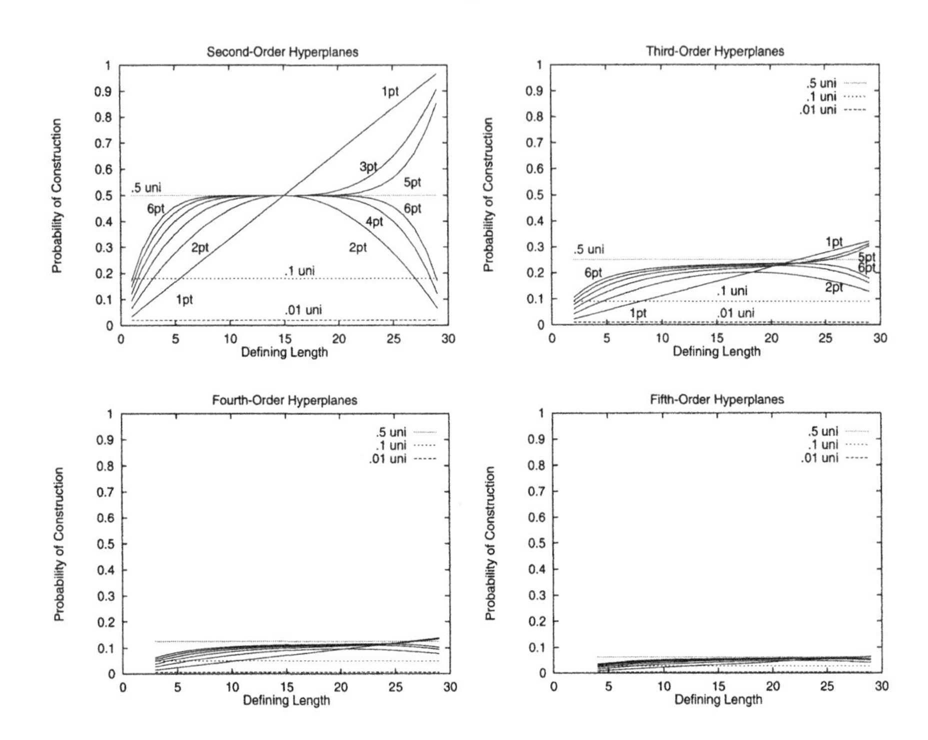

Fig. 4.6. $P_c(H_k)$ of H_2, H_3, H_4, and H_5 when $L = 30$ and $P_{eq} = 0.0$, for P_0 uniform recombination

$$
\begin{aligned}
P_c(H_k, P_0 \mid \mathcal{S}) \; = \; & \\
& P_{s,orig}(H_m, P_0)\; P_{s,other}(H_n, P_0) + P_{s,other}(H_m, P_0)\; P_{s,orig}(H_n, P_0) - \\
& P_{s,both}(H_m, P_0)\; P_{s,both}(H_n, P_0)
\end{aligned}
$$

This equation reflects the decomposition of construction into two independent survival events. The first term is the probability that H_m will survive on the original string, while H_n switches (i.e., both hyperplanes survive on the first offspring). The second term is the probability that H_n will survive on the original string, while H_m switches (i.e., both hyperplanes survive on the other offspring). The third term reflects the joint probability that both hyperplanes survive on both strings, and must be subtracted. If one assumes independence between alleles, the last term is equivalent to:

$$P_{eq}(M)\; P_{eq}(N) \;=\; \prod_{d \in M} P_{eq}(d) \prod_{d \in N} P_{eq}(d) \;=\; \prod_{d \in K} P_{eq}(d)$$

As usual, a further simplification can be made by assuming identicalness ($P_{eq}(d) = P_{eq},\ \forall\ d \in K$). In this case:

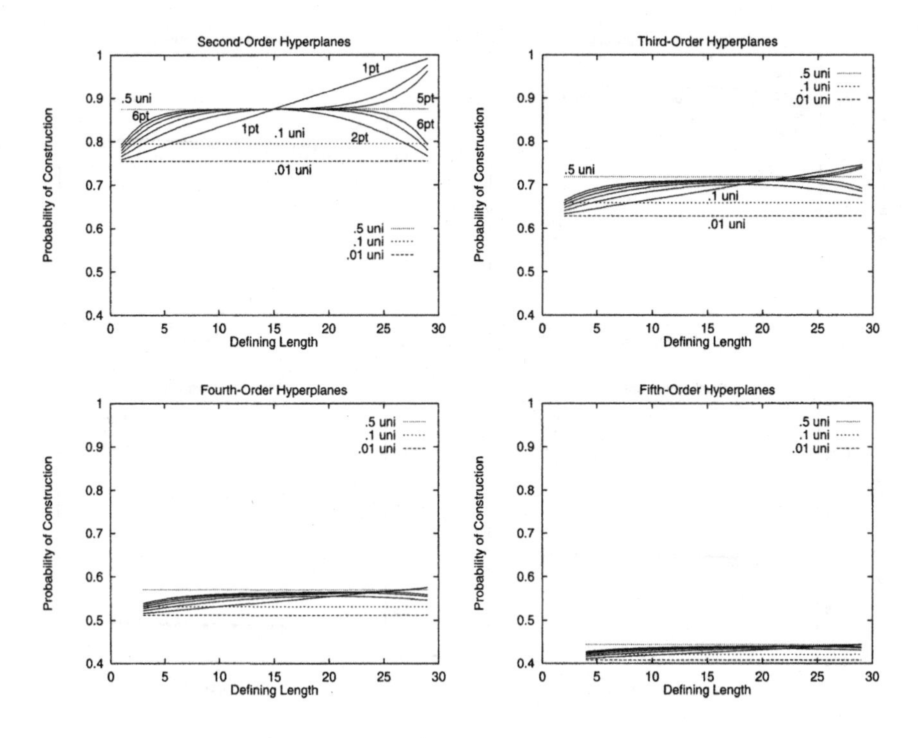

Fig. 4.7. $P_c(H_k)$ of H_2, H_3, H_4, and H_5 when $L = 30$ and $P_{eq} = 0.5$, for P_0 uniform recombination

$$P_{s,orig}(H_k, P_0) = \sum_{x=0}^{k} \binom{k}{x} P_0{}^x (1 - P_0)^{k-x} P_{eq}{}^x$$

$$P_{s,other}(H_k, P_0) = \sum_{x=0}^{k} \binom{k}{x} P_0{}^x (1 - P_0)^{k-x} P_{eq}{}^{k-x}$$

$$P_{s,both}(H_k, P_0) = \sum_{x=0}^{k} \binom{k}{x} P_0{}^x (1 - P_0)^{k-x} P_{eq}{}^k = P_{eq}{}^k$$

Making the two assumptions of independence and identicalness, the probability of construction of hyperplane H_k is:

$$\begin{aligned} P_c(H_k, P_0 \mid \mathcal{S}) = {} & P_{s,orig}(H_m, P_0)\; P_{s,other}(H_n, P_0) + \\ & P_{s,other}(H_m, P_0)\; P_{s,orig}(H_n, P_0) - P_{eq}{}^k \end{aligned} \tag{4.5}$$

Figures 4.6, 4.7, and 4.8 graph the probability of construction $P_c(H_k)$ for both n-point recombination and P_0 uniform recombination, when $P_{eq} = 0.0$, $P_{eq} = 0.5$, and $P_{eq} = 0.75$. The graphs indicate that, as expected, P_0 uniform recombination is only affected by the order k of the hyperplanes – it is not affected by defining lengths. The constructive ability of P_0 uniform

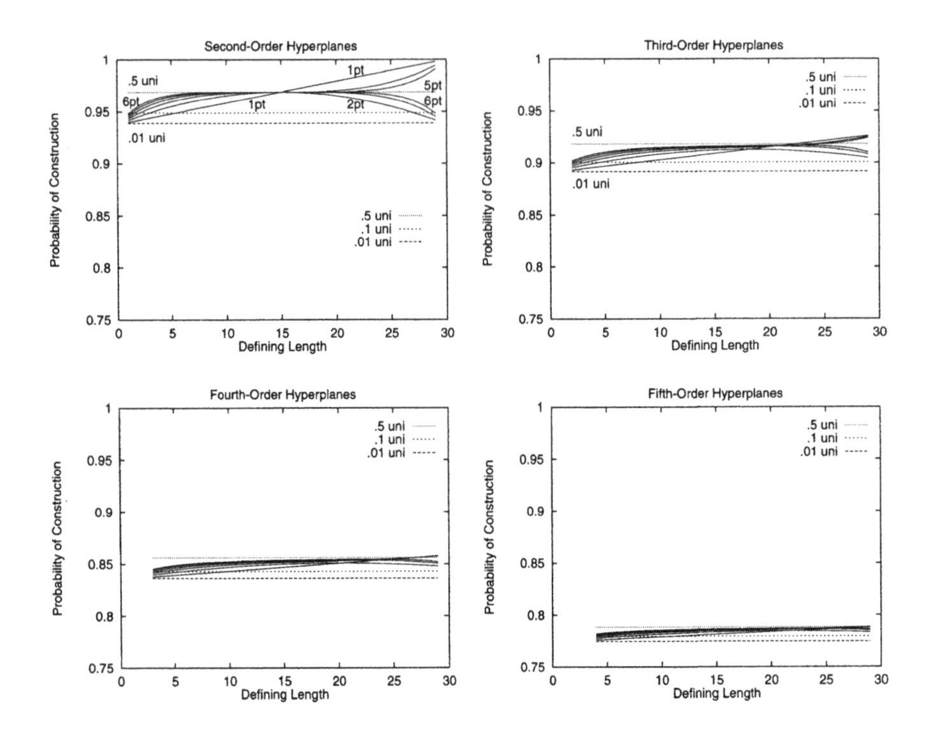

Fig. 4.8. $P_c(H_k)$ of H_2, H_3, H_4, and H_5 when $L = 30$ and $P_{eq} = 0.75$, for P_0 uniform recombination

recombination varies with P_0. The most constructive setting for P_0 uniform recombination is when $P_0 = 0.5$. In general this is more constructive than n-point recombination (especially for higher-order hyperplanes). Finally, the amount of construction caused by P_0 uniform recombination can be lowered by simply reducing P_0 – when $P_0 = 0.0$ there is minimal construction. These *relative* results do not appear to change as the population homogeneity P_{eq} changes.

Finally, once again the graphs for construction are qualitatively the opposite of the graphs for survival given in Chap. 3. This can be seen, for example, by comparing Fig. 3.9 with Fig. 4.6. These results provide strong evidence that more disruptive recombination operators are more likely to construct. Chapter 5 will analyze this relationship in more detail. However, before doing this it is instructive to analyze several special cases of P_0 uniform recombination.

4.4.1 A Special Case: $P_{eq} = 0.0$

As an interesting check of Eq. 4.5, consider the case where there is maximum population diversity ($P_{eq} = 0.0$). In that case:

$$
\begin{aligned}
P_{\text{s,orig}}(H_k, P_0) &= (1 - P_0)^k \\
P_{\text{s,other}}(H_k, P_0) &= {P_0}^k \\
P_{\text{s,both}}(H_k, P_0) &= 0
\end{aligned}
$$

In this simplified case the probability of construction is:

$$P_c(H_k, P_0 \mid S) = (1 - P_0)^m \, {P_0}^n + {P_0}^m \, (1 - P_0)^n \tag{4.6}$$

This indicates that the alleles associated with H_m (or H_n) move to the other individual, while the alleles associated with H_n (or H_m) do not.

4.4.2 A Special Case: $P_0 = 0.0$

As stated earlier in Chap. 3, one nice feature of P_0 uniform recombination is that it can be turned off simply by setting P_0 to 0.0. In that case:

$$
\begin{aligned}
P_{\text{s,orig}}(H_k, P_0 = 0.0) &= 1 \\
P_{\text{s,other}}(H_k, P_0 = 0.0) &= {P_{\text{eq}}}^k \\
P_{\text{s,both}}(H_k, P_0 = 0.0) &= {P_{\text{eq}}}^k
\end{aligned}
$$

In this simplified case Eq. 4.5 (the probability of construction) is:

$$P_c(H_k, P_0 = 0 \mid S) = {P_{\text{eq}}}^n + {P_{\text{eq}}}^m - {P_{\text{eq}}}^k \tag{4.7}$$

which only depends on the population homogeneity and the order of the building blocks H_n and H_m.

4.4.3 A Special Case: $P_0 = 0.5$

Once again another simplification can be made in Eq. 4.5 if $P_0 = 0.5$. In that case each recombination event occurs with probability $1/2^k$ and:

$$
\begin{aligned}
P_{\text{s,orig}}(H_k, P_0 = 0.5) &= \frac{1}{2^k}\left[\sum_{x=0}^{k} \binom{k}{x} {P_{\text{eq}}}^x 1^{k-x}\right] \\
P_{\text{s,other}}(H_k, P_0 = 0.5) &= \frac{1}{2^k}\left[\sum_{x=0}^{k} \binom{k}{x} 1^x {P_{\text{eq}}}^{k-x}\right] \\
P_{\text{s,both}}(H_k, P_0 = 0.5) &= {P_{\text{eq}}}^k
\end{aligned}
$$

Once again the binomial expansion simplifies the equations:

$$
\begin{aligned}
P_{\text{s,orig}}(H_k, P_0 = 0.5) &= \frac{(1 + P_{\text{eq}})^k}{2^k} \\
P_{\text{s,other}}(H_k, P_0 = 0.5) &= \frac{(1 + P_{\text{eq}})^k}{2^k} \\
P_{\text{s,both}}(H_k, P_0 = 0.5) &= {P_{\text{eq}}}^k
\end{aligned}
$$

This will then simplify the probability of construction:

$$
\begin{aligned}
P_c(H_k, P_0 = 0.5 \mid \mathcal{S}) \; = & \\
& P_{s,\text{orig}}(H_m, P_0 = 0.5)\; P_{s,\text{other}}(H_n, P_0 = 0.5) \; + \\
& P_{s,\text{other}}(H_m, P_0 = 0.5)\; P_{s,\text{orig}}(H_n, P_0 = 0.5) \; - \\
& P_{s,\text{both}}(H_m, P_0 = 0.5)\; P_{s,\text{both}}(H_n, P_0 = 0.5)
\end{aligned}
$$

$$
\begin{aligned}
P_c(H_k, P_0 = 0.5 \mid \mathcal{S}) \; = & \\
& \frac{(1+P_{eq})^m}{2^m}\frac{(1+P_{eq})^n}{2^n} + \frac{(1+P_{eq})^m}{2^m}\frac{(1+P_{eq})^n}{2^n} - {P_{eq}}^m {P_{eq}}^n
\end{aligned}
$$

$$
P_c(H_k, P_0 = 0.5 \mid \mathcal{S}) \; = \; \frac{2(1+P_{eq})^k}{2^k} - {P_{eq}}^k
$$

This result is surprising. For 0.5 uniform recombination, the probability of construction of H_k from H_m and H_n does not depend on the *order* of H_m and H_n! This even holds if $m = 0$ or $n = 0$, which corresponds to simple survival. This can be seen by noticing that $P_s(H_k, P_0 = 0.5)$, which was computed in Eq. 3.16 of Chap. 3, has precisely the same value as $P_c(H_k, P_0 = 0.5 \mid \mathcal{S})$. Thus:

$$
P_s(H_k, P_0 = 0.5) \; = \; P_c(H_k, P_0 = 0.5 \mid \mathcal{S}) \quad \forall\, \mathcal{S}
$$

Since this result is extremely nonintuitive, it is useful to consider the following alternative proof. It turns out that this result stems from an interesting relationship between situations and recombination events. Consider the following formulation of the probability of survival:

$$
P_s(H_k) \; = \; P_c(H_k \mid \mathcal{S} = 0) \; = \; \sum_{\mathcal{R}} P(\mathcal{R}) P_c(H_k \mid \mathcal{S} = 0 \wedge \mathcal{R})
$$

Now for 0.5 uniform recombination $P(\mathcal{R}) = 1/2^k \;\; \forall\, \mathcal{R}$, so the probability of survival is:

$$
\begin{aligned}
P_s(H_k, P_0 = 0.5) \; = & \\
P_c(H_k, P_0 = 0.5 \mid \mathcal{S} = 0) \; = & \; \frac{1}{2^k} \sum_{\mathcal{R}} P_c(H_k, P_0 = 0.5 \mid \mathcal{S} = 0 \wedge \mathcal{R})
\end{aligned}
\tag{4.8}
$$

Consider the following formulation of the probability of construction:

$$
P_c(H_k \mid \mathcal{S}) \; = \; \sum_{\mathcal{R}} P(\mathcal{R}) P_c(H_k \mid \mathcal{S} \wedge \mathcal{R})
$$

Again, for 0.5 uniform recombination $P(\mathcal{R}) = 1/2^k \;\; \forall\, \mathcal{R}$, so the probability of construction is:

$$P_c(H_k, P_0 = 0.5 \mid \mathcal{S}) = \frac{1}{2^k} \sum_{\mathcal{R}} P_c(H_k, P_0 = 0.5 \mid \mathcal{S} \wedge \mathcal{R})$$

To complete this proof one has to understand that:

$$P_c(H_k \mid \mathcal{S} \wedge \mathcal{R}) = P_c(H_k \mid \mathcal{S} \oplus z \wedge \mathcal{R} \oplus z) \quad \forall\, z\ \forall\, \mathcal{S}\ \forall\, \mathcal{R} \tag{4.9}$$

The variable z represents any integer and the $\oplus$ operator represents addition modulo 2^k. In other words, since there are 2^k situations and 2^k recombination events, nothing changes if both the situation and the recombination event are changed the same way. For example, suppose one considers the situation $\mathcal{S} = 0$ and recombination event $\mathcal{R} = 0$. The situation $\mathcal{S} = 0$ indicates that all of the alleles for hyperplane H_k are in the first parent. The recombination event $\mathcal{R} = 0$ indicates that no alleles are exchanged during recombination. Now also consider situation $\mathcal{S} = 1$ and recombination event $\mathcal{R} = 1$. In this case the second parent contains one of the desired alleles. However, since $\mathcal{R} = 1$ will in fact exchange that allele, the offspring will be the same as that produced from situation $\mathcal{S} = 0$ and recombination event $\mathcal{R} = 0$. This example is easily generalized to yield Eq. 4.9. This equation is crucial to the investigation of the disruptive and constructive aspects of recombination in Chap. 5. If we let $z = -\mathcal{S}$ (modulo 2^k) we can express the probability of construction for 0.5 uniform recombination as an expression containing only the one situation $\mathcal{S} = 0$:

$$P_c(H_k, P_0 = 0.5 \mid \mathcal{S}) = \frac{1}{2^k} \sum_{\mathcal{R}} P_c(H_k, P_0 = 0.5 \mid \mathcal{S} = 0 \wedge \mathcal{R} \ominus \mathcal{S})$$

The $\ominus$ operator represents subtraction modulo 2^k. Since the summation is summing over all recombination events (they are just shifted by $\mathcal{S}$), this is equivalent to:

$$P_c(H_k, P_0 = 0.5 \mid \mathcal{S}) = \frac{1}{2^k} \sum_{\mathcal{R}} P_c(H_k, P_0 = 0.5 \mid \mathcal{S} = 0 \wedge \mathcal{R})$$

Note that the right-hand side is equal to the probability of survival, $P_s(H_k, P_0 = 0.5)$, as shown in Eq. 4.8, so:

$$P_s(H_k, P_0 = 0.5) = P_c(H_k, P_0 = 0.5 \mid \mathcal{S}) \quad \forall\, \mathcal{S}$$

Thus, by using Eq. 3.16 we get:

$$\begin{aligned} P_s(H_k, P_0 = 0.5) &= \\ P_c(H_k, P_0 = 0.5 \mid \mathcal{S}) &= \frac{2(1 + P_{eq})^k}{2^k} - {P_{eq}}^k \quad \forall\, \mathcal{S} \end{aligned} \tag{4.10}$$

Thus the probability of construction of H_k, under 0.5 uniform recombination, and given some situation $\mathcal{S}$, does not depend on $\mathcal{S}$. From this we can also conclude that the average probability of construction (Eq. 4.4) is also the same:

$$P_c(H_k, P_0 = 0.5) = \frac{1}{2^k - 2} \sum_{S=1}^{2^k-2} P_c(H_k, P_0 = 0.5 \mid S)$$

$$P_c(H_k, P_0 = 0.5) = \frac{2(1 + P_{eq})^k}{2^k} - {P_{eq}}^k \quad (4.11)$$

The key to this behavior has both to do with the mapping between situations and recombination events, as well as the fact that $P(\mathcal{R})$ is the same for all $\mathcal{R}$ with 0.5 uniform recombination.

4.5 Expected Number of Offspring in H_k

Thus far this chapter has provided a thorough analysis of the probability that a hyperplane will be constructed via recombination. However, as stated at the beginning of this chapter, a comparison of recombination with mutation will require the computation of the expected number of offspring in a hyperplane. Thus, the goal now is to convert the probability analysis in this chapter into an analysis yielding the expected number of offspring that will reside in H_k after recombination has occurred.

Once again consider A to be a random variable that describes the number of parents (of the two parents considered for recombination) that are instances of H_k. For survival analysis it is always assumed that at least one parent is an instance of H_k, so A can take on values 1 and 2. However, for construction it is also possible for neither parent to be in H_k, so for construction A can take on values 0, 1, and 2. Let B be a random variable describing the number of offspring that are instances of H_k. B can take on values 0, 1, and 2. We can write:

$$P_c(H_k \mid S) = P(B = 1 \vee 2 \mid A = 0 \vee 1 \vee 2)$$

The expression for $P_c(H_k \mid S)$ can be expanded:

$$\begin{aligned} P_c(H_k \mid S) = \; & P(B = 1 \wedge A = 0) + P(B = 1 \wedge A = 1) + P(B = 1 \wedge A = 2) + \\ & P(B = 2 \wedge A = 0) + P(B = 2 \wedge A = 1) + P(B = 2 \wedge A = 2) \end{aligned}$$

$$\begin{aligned} P_c(H_k \mid S) = \; & P(B = 1 \mid A = 0)P(A = 0) + P(B = 1 \mid A = 1)P(A = 1) + \\ & P(B = 1 \mid A = 2)P(A = 2) + P(B = 2 \mid A = 0)P(A = 0) + \\ & P(B = 2 \mid A = 1)P(A = 1) + P(B = 2 \mid A = 2)P(A = 2) \end{aligned}$$

Once again several terms can be removed (the third, fourth, and fifth), since it is impossible for recombination to construct only one offspring in H_k if both parents are in H_k, and it is also impossible to construct two offspring in H_k if less than two parents are in H_k. Finally if both parents are in H_k, both offspring must be in H_k. Thus we get:

$$P_c(H_k \mid \mathcal{S}) = \\ P(B=1 \mid A=0)P(A=0) + P(B=1 \mid A=1)P(A=1) + P(A=2)$$

Now the expected number of offspring in H_k can be computed as follows:

$$E[B \mid \mathcal{S}] = \sum_{b \in \{0,1,2\}} b \times P(B=b) = P(B=1) + 2P(B=2)$$

For the sake of clarity, denote $E[B \mid \mathcal{S}]$ to be $E_c[B_k \mid \mathcal{S}]$. $E_c[B_k \mid \mathcal{S}]$ will refer to the expected number of offspring that will be in H_k. The 'c' subscript is a reminder that the situation is one in which the construction of an individual in H_k is at stake.

$$\begin{aligned} E_c[B_k \mid \mathcal{S}] &= \\ & P(B=1 \mid A=0)P(A=0) + P(B=1 \mid A=1)P(A=1) + \\ & P(B=1 \mid A=2)P(A=2) + 2P(B=2 \mid A=0)P(A=0) + \\ & 2P(B=2 \mid A=1)P(A=1) + 2P(B=2 \mid A=2)P(A=2) \end{aligned}$$

Again, the third, fourth, and fifth terms disappear:

$$E_c[B_k \mid \mathcal{S}] = \\ P(B=1 \mid A=0)P(A=0) + P(B=1 \mid A=1)P(A=1) + 2P(A=2)$$

Thus:

$$E_c[B_k \mid \mathcal{S}] = P_c(H_k \mid \mathcal{S}) + P(A=2)$$

Now the probability that two parents are in H_k is controlled by P_{eq} (assuming independence and identicalness, as stated earlier). Clearly $P(A = 2) = {P_{eq}}^k$, so:

$$E_c[B_k \mid \mathcal{S}] = P_c(H_k \mid \mathcal{S}) + {P_{eq}}^k \tag{4.12}$$

Finally we can compute the average $E_c[B_k]$ over the $2^k - 2$ constructive situations:

$$E_c[B_k] = \frac{1}{2^k - 2} \sum_{\mathcal{S}=1}^{2^k-2} E_c[B_k \mid \mathcal{S}]$$

$$E_c[B_k] = \frac{1}{2^k - 2} \sum_{\mathcal{S}=1}^{2^k-2} [P_c(H_k \mid \mathcal{S}) + {P_{eq}}^k]$$

$$E_c[B_k] = P_c(H_k) + {P_{eq}}^k \tag{4.13}$$

So, we can see that the computation of the expected number of offspring that will reside in H_k after construction by recombination is easily computed from $P_c(H_k)$ and the population homogeneity P_{eq}. Notice that if $P_{eq} = 0.0$, $E_c[B_k] = P_c(H_k)$, which is what we would expect. Notice also that as P_{eq} approaches one both $P_c(H_k)$ and ${P_{eq}}^k$ approach one, so $E_c[B_k]$ approaches two as we would expect.

For a particular hyperplane H_k and population homogeneity P_{eq}, the previous construction graphs in this chapter would only need to be offset by ${P_{eq}}^k$ to produce graphs for $E_c[B_k]$. Since the offset would be the same for each recombination operator within each graph, the relationships between the different recombination operators would not change.

4.6 Summary

This chapter first computes the probability $P_c(H_k)$ that a kth-order hyperplane H_k will be constructed via recombination, given that one parent is a member of a lower-order hyperplane H_m and that the other parent is a member of another lower-order hyperplane H_n. We restrict the situation such that the two lower-order hyperplanes H_m and H_n are nonoverlapping, and $k = m + n$.

Both n-point and P_0 uniform recombination are analyzed. The cardinality of the alphabet and the population homogeneity are taken into account in a natural fashion. The results indicate that the constructive aspect of n-point recombination is affected by both the defining lengths and the order k of the hyperplane, while P_0 uniform recombination is only affected by the order. All forms of n-point recombination are reasonably nonconstructive of short schemata, but the higher n is, the greater the construction. For long schemata, n-point recombination when n is odd is more constructive than when n is even.

The constructive ability of P_0 uniform recombination varies with P_0. The most constructive setting for P_0 uniform recombination is when $P_0 = 0.5$. In general this is more constructive than n-point recombination (especially for higher-order hyperplanes). Finally, the amount of construction caused by P_0 uniform recombination can be lowered by simply reducing P_0 – when $P_0 = 0.0$ there is minimal construction. These relative results do not appear to change as the population homogeneity P_{eq} changes.

This chapter concludes by computing the expected number of offspring $E_c[B_k]$ that reside in H_k after construction via recombination, and shows that this is a simple function of $P_c(H_k)$, the population homogeneity P_{eq}, and the order of the hyperplane k. The computation of $E_c[B_k]$ is important because it will allow for a fair comparison with mutation (in Chaps. 6–8).

A comparison of the graphs from this chapter and Chap. 3 illustrate that the graphs for construction are qualitatively the opposite of the graphs for survival given in Chap. 3. These results provide strong evidence that more disruptive recombination operators are more likely to construct (and vice versa). The natural question is whether this in fact holds in general. Chapter 5 addresses this question in detail and shows that a form of No Free Lunch theorem (Wolpert and Macready 1995) holds for construction and survival for recombination operators.

5. Survival and Construction Schema Theory for Recombination

5.1 Introduction

Chapter 3 provided an analysis of how likely it is to disrupt hyperplanes via recombination. Chapter 4 considered the more positive view of recombination as an operator than can construct hyperplanes from lower-order hyperplanes. One interesting observation can be made by comparing the graphs for disruption in Chap. 3 to those for construction in Chap. 4 – namely, that the graphs are qualitatively related in the sense that if one recombination operator is more disruptive, it is also more constructive. The results are intuitively plausible and hold for every set of graphs we have been able to generate.

The natural question is whether this qualitative relationship does in fact hold in general. Furthermore, is the relationship not only qualitative, but quantitative? In other words, if one knows the probability that a given recombination operator will disrupt a given hyperplane, can one immediately compute the probability that the given recombination operator will construct that hyperplane? These questions are investigated in this chapter.

5.2 Survival and Construction

Suppose one considers both the survival and construction graphs and combines them to give the probability of construction or survival. This can be done by considering all 2^k situations $\mathcal{S}$ (as defined in Chap. 4):

$$P_{\mathrm{c,s}}(H_k) = \frac{1}{2^k} \sum_{\mathcal{S}=0}^{2^k-1} P_{\mathrm{c,s}}(H_k \mid \mathcal{S}) \tag{5.1}$$

$P_{\mathrm{c,s}}(H_k)$ represents the probability that either offspring of recombination will reside in H_k, either via survival or construction. The notation $P_{\mathrm{c,s}}(H_k \mid \mathcal{S})$ refers to $P_{\mathrm{s}}(H_k)$ when $\mathcal{S} = 0$ or $\mathcal{S} = 2^k - 1$ (i.e., when there is really no construction but only survival). Otherwise it refers to $P_{\mathrm{c}}(H_k \mid \mathcal{S})$, which is an actual construction.[1] Thus expanding the above equation yields:

[1] The probability of survival $P_{\mathrm{s}}(H_k)$ is defined in Chap. 3 while the probability of construction $P_{\mathrm{c}}(H_k \mid \mathcal{S})$ is defined in Chap. 4.

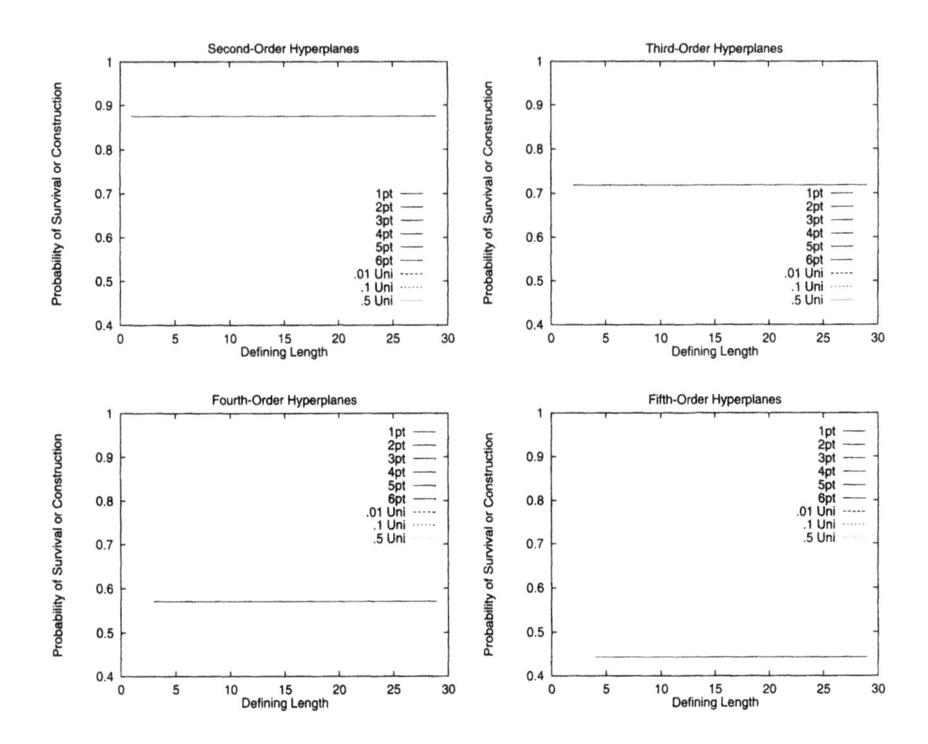

Fig. 5.1. $P_{c,s}(H_k)$ of H_2, H_3, H_4, and H_5 when $L = 30$ and $P_{eq} = 0.5$. The results are the same regardless of recombination operator.

$$P_{c,s}(H_k) = \frac{1}{2^k}\left[P_{c,s}(H_k \mid \mathcal{S}=0) + P_{c,s}(H_k \mid \mathcal{S}=2^k-1) + \sum_{\mathcal{S}=1}^{2^k-2} P_{c,s}(H_k \mid \mathcal{S})\right]$$

But the first two terms are simply the probability of survival (computed in Chap. 3), while the summation is proportional to the average probability of construction (see Eq. 4.4 in Chap. 4). Thus:

$$P_{c,s}(H_k) = \frac{1}{2^k}\left[2P_s(H_k) + (2^k-2)P_c(H_k)\right] \tag{5.2}$$

This equation illustrates how to compute $P_{c,s}(H_k)$ by averaging the disruption (survival) results from Chap. 3 and the construction results from Chap. 4. Surprisingly, when $P_{c,s}(H_k)$ is computed in this fashion, the results are the same regardless of the recombination operator! Figure 5.1 illustrates this for $P_{eq} = 0.5$, for a wide range of n-point and P_0 uniform recombination operators. Similar results are obtained with $P_{eq} = 0.0$ and $P_{eq} = 0.75$ (see Table 5.1).

This chapter will prove that this is true in general. The key point is that the probability of survival or construction, when averaged over all situations,

Table 5.1. $P_{c,s}(H_k)$ values obtained by averaging $P_s(H_k)$ and $P_c(H_k)$. The results are the same regardless of recombination operator.

	$P_{eq} = 0.0$	$P_{eq} = 0.5$	$P_{eq} = 0.75$
H_2	0.5	0.875	0.968750
H_3	0.25	0.718750	0.917969
H_4	0.125	0.570312	0.855957
H_5	0.0625	0.443359	0.788513

does *not* depend on the form of recombination. It only depends on the population homogeneity.

Consider a breakdown of $P_{c,s}(H_k)$ over all situations $\mathcal{S}$ and recombination events $\mathcal{R}$ as follows:

$$P_{c,s}(H_k) = \frac{1}{2^k} \sum_{\mathcal{S}} P_{c,s}(H_k \mid \mathcal{S})$$

$$P_{c,s}(H_k) = \frac{1}{2^k} \sum_{\mathcal{R}} \sum_{\mathcal{S}} P(\mathcal{R}) \; P_{c,s}(H_k \mid \mathcal{S} \wedge \mathcal{R})$$

$$P_{c,s}(H_k) = \frac{1}{2^k} \sum_{\mathcal{R}} \left[P(\mathcal{R}) \sum_{\mathcal{S}} P_{c,s}(H_k \mid \mathcal{S} \wedge \mathcal{R}) \right]$$

Chapter 4 introduced the following important fact that illustrates a tight relationship between situations and recombination events:

$$P_{c,s}(H_k \mid \mathcal{S} \wedge \mathcal{R}) = P_{c,s}(H_k \mid \mathcal{S} \oplus z \wedge \mathcal{R} \oplus z) \quad \forall z \;\; \forall \mathcal{S} \;\; \forall \mathcal{R} \tag{5.3}$$

The variable z represents any integer and the $\oplus$ operator represents addition modulo 2^k. Since this is an important relationship we repeat the discussion in Chap. 4. The key point is that since there are 2^k situations and 2^k recombination events, nothing changes if both the situation and the recombination event are changed the same way. For example, suppose one considers the situation $\mathcal{S} = 0$ and recombination event $\mathcal{R} = 0$. The situation $\mathcal{S} = 0$ indicates that all of the alleles for hyperplane H_k are in the first parent. The recombination event $\mathcal{R} = 0$ indicates that no alleles are exchanged during recombination. Now also consider situation $\mathcal{S} = 1$ and recombination event $\mathcal{R} = 1$. In this case the second parent contains one of the desired alleles. However, since $\mathcal{R} = 1$ will in fact exchange that allele, the offspring will be the same as that produced from situation $\mathcal{S} = 0$ and recombination event $\mathcal{R} = 0$. This example is easily generalized to yield Eq. 5.3. If we let $z = -\mathcal{R}$ (modulo 2^k) we can rephrase the inner sum in terms of one recombination event only:

$$P_{c,s}(H_k) = \frac{1}{2^k}\sum_{\mathcal{R}}\left[P(\mathcal{R})\sum_{\mathcal{S}} P_{c,s}(H_k \mid \mathcal{S} \ominus \mathcal{R} \wedge \mathcal{R} = 0)\right]$$

The $\ominus$ operator represents subtraction modulo 2^k. Since the inner summation is summing over all situations (they are just shifted by $\mathcal{R}$), this is equivalent to:

$$P_{c,s}(H_k) = \frac{1}{2^k}\sum_{\mathcal{R}}\left[P(\mathcal{R})\sum_{\mathcal{S}} P_{c,s}(H_k \mid \mathcal{S} \wedge \mathcal{R} = 0)\right]$$

This inner summation can now be separated from the events $\mathcal{R}$:

$$P_{c,s}(H_k) = \frac{1}{2^k}\left[\sum_{\mathcal{S}} P_{c,s}(H_k \mid \mathcal{S} \wedge \mathcal{R} = 0)\right]\left[\sum_{\mathcal{R}} P(\mathcal{R})\right]$$

Now, the probability of all recombination events must sum to 1.0, so:

$$P_{c,s}(H_k) = \frac{1}{2^k}\left[\sum_{\mathcal{S}} P_{c,s}(H_k \mid \mathcal{S} \wedge \mathcal{R} = 0)\right] \quad (5.4)$$

Clearly this does not depend on the form of recombination, since the probability of recombination events is absent. What this says is that $P_{c,s}(H_k)$ is the same, regardless of the form of recombination. It is important to note that this is a general result; no assumptions concerning independence and identicalness (see Chap. 3 or 4) have been made. Such assumptions are only necessary if one wants to compute a value for $P_{c,s}(H_k)$, as follows.

Recall Equation 5.1:

$$P_{c,s}(H_k) = \frac{1}{2^k}\sum_{\mathcal{S}=0}^{2^k-1} P_{c,s}(H_k \mid \mathcal{S})$$

Since Eq. 5.4 has shown that the form of recombination is irrelevant, it suffices to use the computations for 0.5 uniform recombination from Chap. 4 to compute Eq. 5.1. By making the standard independence and identicalness assumptions (see Chap. 3 or 4 for details) Eq. 4.10 proved that:

$$P_{c,s}(H_k, P_0 = 0.5 \mid \mathcal{S}) = \frac{2(1+P_{eq})^k}{2^k} - {P_{eq}}^k \quad \forall\, \mathcal{S}$$

Thus, for any recombination operator:

$$P_{c,s}(H_k) = \frac{2(1+P_{eq})^k}{2^k} - {P_{eq}}^k \quad (5.5)$$

Table 5.2 gives the computationally derived values, using this equation. The agreement with the empirically derived averaged results (see Table 5.1)

Table 5.2. $P_{c,s}(H_k)$ values obtained by theory

	$P_{eq} = 0.0$	$P_{eq} = 0.5$	$P_{eq} = 0.75$
H_2	0.5	0.875	0.968750
H_3	0.25	0.718750	0.917969
H_4	0.125	0.570312	0.855957
H_5	0.0625	0.443359	0.788513

is perfect. This agreement provides nice confirmation of the math and code used to generate the curves for the graphs in Chaps. 3 and 4.

More importantly, however, is that this result indicates that disruption and construction are tightly related. Consider Eq. 5.2 for the probability of construction or survival:

$$P_{c,s}(H_k) = \frac{1}{2^k}\left[2P_s(H_k) + (2^k - 2)P_c(H_k)\right] = \frac{2(1+P_{eq})^k}{2^k} - {P_{eq}}^k$$

Thus, any decrease in disruption (which is an increase in survival), *must* be countered by a decrease in construction and vice versa. Disruption and construction are not only related qualitatively – they are related quantitatively. Since the relationship is quantitative, it also is useful computationally. Calculating $P_c(H_k)$ involves a fair amount of computation (due to the large number of situations $\mathcal{S}$) – far more than the calculation of $P_s(H_k)$ and $P_{c,s}(H_k)$. The above equation indicates that $P_c(H_k)$ can be efficiently derived from $P_s(H_k)$ (which depends on the form of recombination and P_{eq}) and $P_{c,s}(H_k)$ (which depends only on P_{eq}).

What we have shown is essentially a No Free Lunch theorem with respect to the disruption (survival) and construction aspects of recombination operators. When averaged over all possible situations, all recombination operators are equivalent. Any loss in disruption (gain in survival) is offset by a loss in construction.

5.3 Expected Number of Offspring in H_k

Thus far this chapter has provided a thorough analysis of the probability that a hyperplane will be constructed or survive via recombination. The goal now is to convert that into a computation yielding the expected number of offspring that will reside in H_k.

For the sake of clarity, let $E_{c,s}[B_k]$ denote the expected number of offspring that will be in H_k. The subscript is a reminder that the situation is one in which the survival or construction of an individual in H_k is at stake (and that we are considering all 2^k situations). Thus, by definition:

$$E_{c,s}[B_k] = \frac{1}{2^k}\sum_{\mathcal{S}=0}^{2^k-1} E_{c,s}[B_k \mid \mathcal{S}]$$

However, Eq. 3.17 in Chap. 3 showed that for the survival situations (after making the independence and identicalness assumptions):

$$E_{c,s}[B_k \mid \mathcal{S} = 0] \;=\; E_{c,s}[B_k \mid \mathcal{S} = 2^k - 1] \;=\; E_s[B_k] \;=\; P_s(H_k) \;+\; {P_{eq}}^k$$

while for the construction situations ($0 < \mathcal{S} < 2^k - 1$), Eq. 4.12 in Chap. 4 showed that:

$$E_{c,s}[B_k \mid \mathcal{S}] \;=\; E_c[B_k \mid \mathcal{S}] \;=\; P_c(H_k \mid \mathcal{S}) \;+\; {P_{eq}}^k$$

Thus:

$$E_{c,s}[B_k] \;=\; \frac{1}{2^k}\sum_{\mathcal{S}=0}^{2^k-1} [P_{c,s}(H_k \mid \mathcal{S}) \;+\; {P_{eq}}^k]$$

$$E_{c,s}[B_k] \;=\; P_{c,s}(H_k) \;+\; {P_{eq}}^k$$

By Equation 5.5:

$$E_{c,s}[B_k] \;=\; \frac{2(1 + P_{eq})^k}{2^k} \tag{5.6}$$

So, we can see that the expected number of offspring that will reside in H_k after survival or construction by recombination is simply a constant for all recombination operators. The constant is determined by the order k and the population homogeneity P_{eq}.

Notice that if $P_{eq} = 0.0$, $E_{c,s}[B_k] = P_{c,s}(H_k)$, which is what we would expect. In fact, in that case:

$$E_{c,s}[B_k] \;=\; \frac{1}{2^{k-1}} \tag{5.7}$$

Notice also that as P_{eq} approaches one, both individuals are increasingly likely to be in H_k, so $E_{c,s}[B_k]$ approaches two as we would expect.

5.4 Summary

This chapter has proven a No Free Lunch theorem with respect to the disruption (survival) and construction aspects of recombination operators. When averaged over all possible situations, all recombination operators are equivalent. Any loss in disruption (gain in survival) is offset by a loss in construction. This relationship is not only qualitative, but quantitative.

Chapter 3, Chapter 4, and this chapter have provided an extensive development of a schema theory for recombination. Since the focus of this book is also on mutation (in order to compare recombination and mutation), the next several chapters will focus on developing a schema theory for mutation.

6. A Survival Schema Theory for Mutation

6.1 Introduction

Chapter 3 computed the probability $P_s(H_k)$ that a kth-order hyperplane H_k will survive recombination, given that one parent is a member of H_k and that the other parent is arbitrary. That chapter concluded by computing the expected number of offspring $E_s[B_k]$ that reside in H_k after recombination, and showed that this is a simple function of $P_s(H_k)$, the population homogeneity P_{eq}, and the order of the hyperplane k.

The goal of this chapter is to provide a similar computation for mutation. Mutation will work on alphabets of cardinality C in the following fashion. An allele is picked for mutation with probability μ (the "mutation rate"). Then that allele is changed to one of the other $C-1$ alleles, uniformly randomly.[1] Mutation is performed independently to both parents, since in almost all EAs mutation is applied independently to every individual in the population.

We omit the intermediary step of computing $P_s(H_k)$ for mutation, due to some differences between mutation and recombination. As stated in Chap. 3, in order to take population homogeneity into account, mutation needs to be considered as a two-parent operator (since population homogeneity is a relationship between multiple individuals in a population). Now, recall that if only one parent is a member of a hyperplane H_k, then survival via recombination means that only one offspring can be a member of H_k. However, survival via mutation means that one or both offspring may be a member of H_k (due to the independence assumption). This renders the computation of $P_s(H_k)$ worthless as a comparison between the two operators, because it obscures information concerning the expected number of offspring that are in H_k. Instead, in this chapter we will compute $E_s[B_k]$ directly.

To summarize – in order to provide for a fair comparison between recombination and mutation, mutation will be treated as if it were a two-parent operator that produces two offspring. Precisely the same random experiment will be performed as with recombination. This chapter will then use this framework to compute the expected number of offspring $E_s[B_k]$ that are in

[1] This form of mutation is reasonable for discrete representations; however, it should be modified for real-valued representations.

a hyperplane H_k, given that one parent is in H_k and the other parent is arbitrary, after mutation has changed the parents.

6.2 Framework

Assume that the following random experiment is being performed. One is given two parents, and one parent is in the schema H_k, while the other parent is an arbitrary string (which may or may not be in the schema H_k). Figure 6.1 provides a pictorial example. The two parents are labeled P1 and P2. P1 is a member of a particular third-order hyperplane H_3, which has defining positions d_1, d_2, and d_3. P2 is some other arbitrary individual.

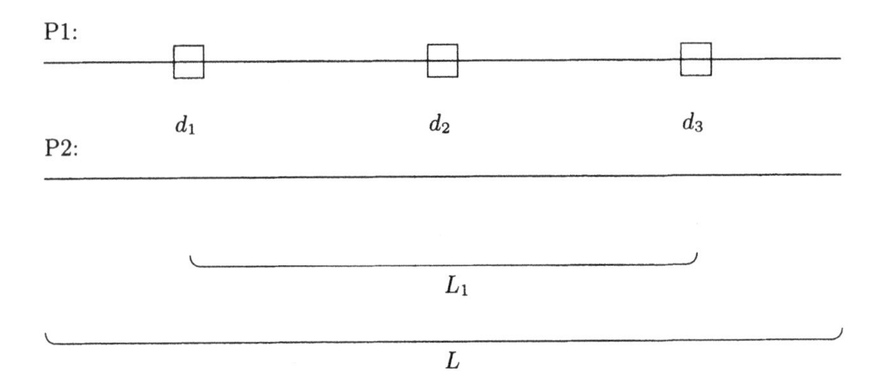

Fig. 6.1. The setup for the random experiment to be performed. P1 is a member of a third-order hyperplane, and P2 is an arbitrary string. Mutation is performed on both parents, producing two offspring.

The random experiment consists of performing mutation on these two parents, producing two children. Schema H_3 (or H_k in general) can either survive or be disrupted. A schema survives if either offspring is in H_3 (H_k), and it is disrupted if neither offspring is in H_3 (H_k). This is the same definition of survival and disruption that was presented for recombination in the earlier chapters.

As before, let B be a random variable describing the number of offspring that are instances of H_k. B can take on values 0, 1, and 2. We can write:

$$E[B] = \sum_{b \in \{0,1,2\}} b \times P(B = b) = P(B = 1) + 2P(B = 2)$$

For the sake of clarity, denote $E[B]$ to be $E_s[B_k, \mu]$. $E_s[B_k, \mu]$ will refer to the expected number of offspring that will be in H_k after μ mutation has been performed. The subscript 's' is a reminder that the experiment being performed is the same as that performed for the survival analysis under recombination.

Without loss of generality, assume that the first parent is in H_k, while the second parent is arbitrary. To compute $E_s[B_k, \mu]$ it is convenient to let Q be a random variable that describes the set of alleles (at the defining positions) in the second individual that do not match H_k. Then we can write $E_s[B_k, \mu]$ as follows:

$$E_s[B_k, \mu] = \sum_Q P(Q)[\, P(B=1 \mid Q) + 2P(B=2 \mid Q)\,]$$

It will be noted that $P(Q)$ depends on the population homogeneity. Deriving precise expressions for the values of $P(Q)$ at a particular point in time is difficult in general since they vary from generation to generation in complex, nonlinear, and interacting ways. We can, however, get considerable insight into the effects of shared alleles on disruption analysis by making simplifying assumptions.

First, as opposed to concentrating on sets, let Q be a random variable that describes the *number* of alleles (at the defining positions) in the second individual that do not match H_k. Q can take on the integer values from 0 to k. For example, if the first parent is a member of the third-order hyperplane #AAA# and the second parent is a member of #ABA#, then Q is 1. Thus we can rewrite $E_s[B_k, \mu]$ as follows:

$$E_s[B_k, \mu] = \sum_{q=0}^{k} P(Q=q)[\, P(B=1 \mid Q=q) + 2P(B=2 \mid Q=q)\,]$$

Let $P_{eq}(d)$ represent the probability that both parents have the same allele at a particular defining position d. Then further assume that $P_{eq}(d)$ is roughly the same for all the defining positions ($P_{eq}(d) = P_{eq}, \ \forall\ d$).[2] Then $P(Q=q)$ is simply:

$$P(Q=q) = \binom{k}{q} (1 - P_{eq})^q \, {P_{eq}}^{k-q}$$

Now consider the derivation of the other terms of $E_s[B_k, \mu]$. In order to have both offspring be in H_k (i.e., $B = 2$), the k alleles in the first parent (associated with the hyperplane H_k) must not be mutated, since the first parent is already in H_k. However, the Q differing alleles in the second parent must be mutated, while the remaining $k - Q$ alleles in the second parent must not be mutated, in order to place the second offspring in H_k. Assuming that one is interested in the hyperplane #AAA# (in the previous example), the As in the two parents must not be mutated, while the B must be mutated

[2] The same assumptions were made in the recombination analysis.

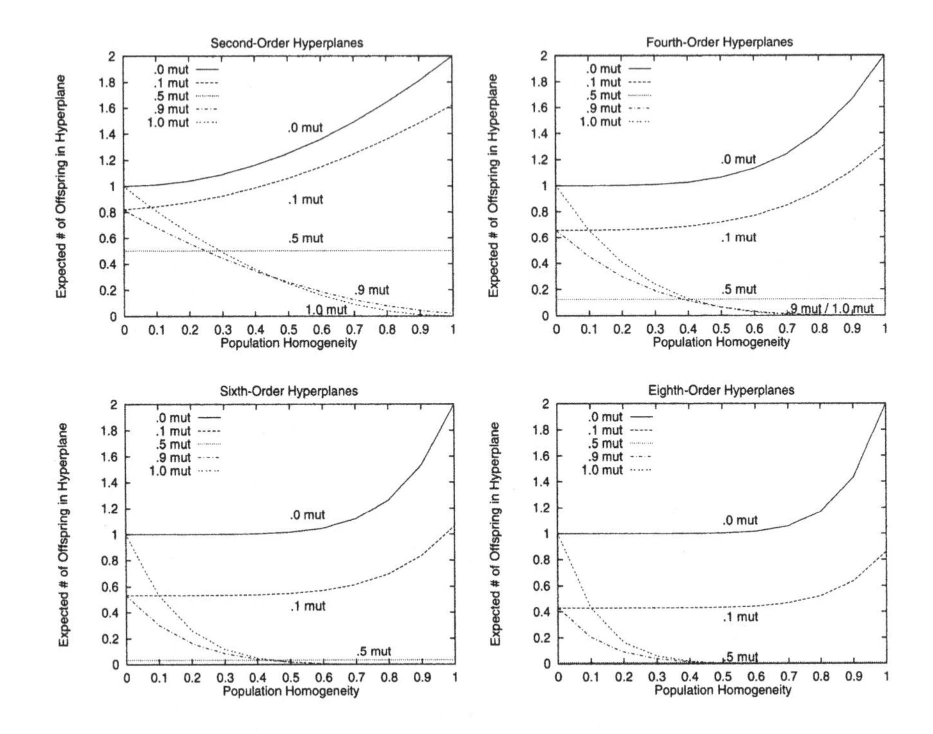

Fig. 6.2. $E_s[B_k, \mu]$ of H_2, H_4, H_6, and H_8 for mutation when $C = 2$

to an A. For a general alphabet of cardinality C, if an allele is mutated, there is a $1/(C-1)$ probability of mutating it to the desired allele. Thus, the probability of placing *both* offspring in H_k is simply computed as:

$$P(B = 2 \mid Q = q) \ = \ (1-\mu)^k \ [\ \left(\frac{\mu}{C-1}\right)^q \ (1-\mu)^{k-q} \ ]$$

where the probability of not mutating the k alleles of the first parent is $(1-\mu)^k$, and the remainder of the expression is the probability of mutating the second parent into the hyperplane H_k.

It is now possible to compute the probability that *only one* offspring will be in H_k. Clearly that will occur if the first parent is kept in H_k while the second parent is not mutated into H_k, or if the first parent is mutated out of H_k while the second parent is mutated into H_k. This can be simply computed by using the components of the previous equation:

$$P(B = 1 \mid Q = q) \ =$$
$$(1-\mu)^k \ [\ 1 \ - \ \left(\frac{\mu}{C-1}\right)^q \ (1-\mu)^{k-q} \ ] \ +$$
$$[\ 1 \ - \ (1-\mu)^k \ ] \ [\ \left(\frac{\mu}{C-1}\right)^q \ (1-\mu)^{k-q} \ ]$$

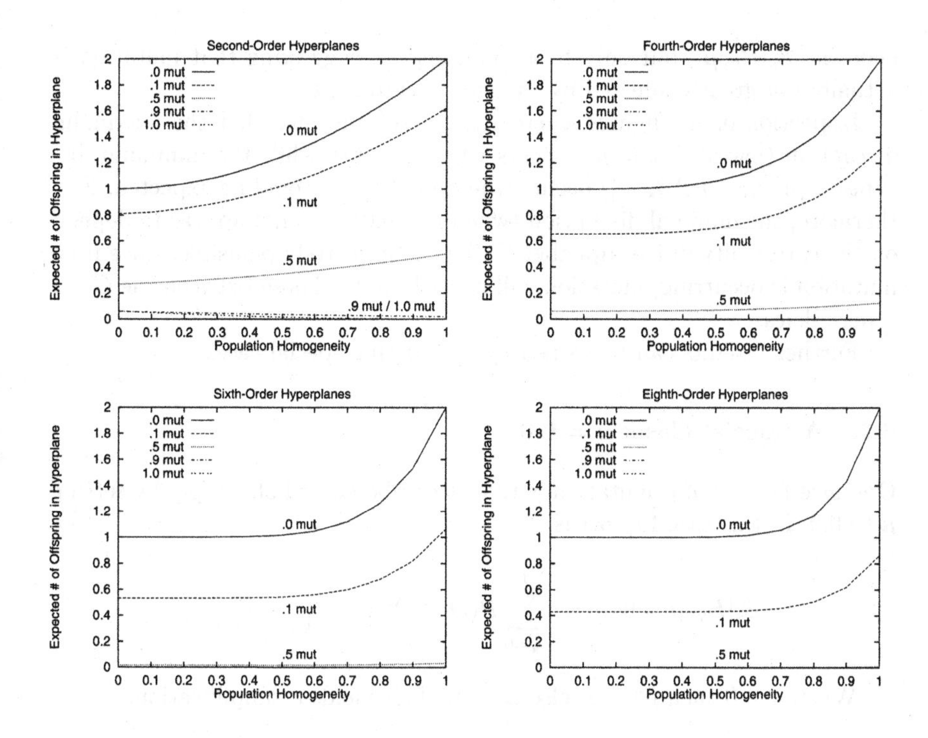

Fig. 6.3. $E_s[B_k, \mu]$ of H_2, H_4, H_6, and H_8 for mutation when $C = 5$

With some simplification $E_s[B_k, \mu]$ can now be expressed for μ mutation:

$$E_s[B_k, \mu] = \tag{6.1}$$
$$\sum_{q=0}^{k} \binom{k}{q} (1 - P_{eq})^q \, P_{eq}{}^{k-q} [\, (1-\mu)^k \, + \left(\frac{\mu}{C-1}\right)^q (1-\mu)^{k-q} \,]$$

Figure 6.2 illustrates $E_s[B_k, \mu]$ when $C = 2$, for mutation rates ranging from 0.0 to 1.0, for $k \in \{2, 4, 6, 8\}$, while P_{eq} ranges from 0.0 to 1.0. Figure 6.3 illustrates the expected number of offspring in H_k when $C = 5$. For a randomly initialized population any allele has probability $1/C$ of being the same as any other allele, so the minimum P_{eq} is simply $1/C$. Since the populations of traditional EAs tend to become more homogeneous with time, it is reasonable to examine only those situations where $P_{eq} > 1/C$.[3]

A number of observations can be made from the figures. The expected number of offspring surviving in H_k ($E_s[B_k, \mu]$) decreases as the order k of the hyperplane increases. $E_s[B_k, \mu]$ increases as P_{eq} increases, for values of μ less than 0.5 (higher values are very rare in practice). Both of the observations are intuitively reasonable. More interestingly, increasing the cardinality C does

[3] However, it is possible that speciating EAs (see Chap. 2) may have $P_{eq} < 1/C$.

decrease $E_s[B_k, \mu]$, but only to a small degree, suggesting that mutation is actually not greatly affected by changes in cardinality.

Inspection of the figures where $P_{eq} > 1/C$ indicates that the maximum disruption (lowest $E_s[B_k, \mu]$) occurs when $\mu = 1.0$, while the minimum disruption (highest $E_s[B_k, \mu]$) occurs when $\mu = 0.0$, as would be expected. Furthermore, the minimal disruption (when $\mu = 0.0$) does not appear to depend on the cardinality of the alphabet C. This is intuitively plausible, since if no mutation is occurring, mutation will never have to choose between the $C - 1$ other alleles.

Further insights can be gained by considering special cases.

6.2.1 A Special Case: $\mu = 0.0$

One nice feature of μ mutation is that it can be turned off simply by setting μ to 0.0. In this case Eq. 6.1 is:

$$E_s[B_k, \mu = 0.0] = \sum_{q=0}^{k} P(Q = q)[\ 1 + \left(\frac{\mu}{C-1}\right)^q\ ]$$

When $q = 0$ then $\mu^q = 1$; else $\mu^q = 0$. This yields a simplification:

$$E_s[B_k, \mu = 0.0] = 2\ P(Q = 0) + \sum_{q=1}^{k} P(Q = q)$$

$$E_s[B_k, \mu = 0.0] = 2\ P(Q = 0) + [\ 1 - P(Q = 0)\ ]$$

$$E_s[B_k, \mu = 0.0] = 1 + P(Q = 0)$$

But $P(Q = 0) = {P_{eq}}^k$, so:

$$E_s[B_k, \mu = 0.0] = 1 + {P_{eq}}^k \tag{6.2}$$

As we would presume, the expected number of offspring in H_k depends only on the population homogeneity P_{eq} and the order k of the hyperplane. This is reasonable, since mutation is in fact turned off. Clearly, C has no effect on this behavior. Thus the maximum likelihood of survival for a hyperplane is when mutation is not run at all, which is intuitively clear. Also, one can see that $E_s[B_k, \mu = 0.0]$ ranges from 1.0 to 2.0 as P_{eq} ranges from 0.0 to 1.0.

6.2.2 A Special Case: $\mu = 0.5$

Let's consider the case when $\mu = 0.5$. Then Eq. 6.1 is:

$$E_s[B_k, \mu = 0.5] \;=\; \sum_{q=0}^{k} P(Q = q)[\; 0.5^k \;+\; \left(\frac{0.5}{C-1}\right)^q \; 0.5^{k-q}\;]$$

$$E_s[B_k, \mu = 0.5] \;=\; \sum_{q=0}^{k} P(Q = q)[\; 0.5^k \;+\; \frac{0.5^k}{(C-1)^q}\;]$$

$$E_s[B_k, \mu = 0.5] \;=\; 0.5^k \;+\; 0.5^k \sum_{q=0}^{k} \frac{P(Q = q)}{(C-1)^q}$$

For $C = 2$ this collapses to:

$$E_s[B_k, \mu = 0.5] \;=\; 0.5^k \;+\; 0.5^k \;=\; \frac{1}{2^{k-1}}$$

Note how $E_s[B_k, \mu = 0.5]$ does not depend on P_{eq} when $C = 2$, but does when $C = 5$ (compare Fig. 6.2 with Fig. 6.3). This makes sense, since with a mutation rate of 0.5 and $C = 2$, the offspring are always just randomly reinitialized individuals, no matter what the parents are. This is not true when $C > 2$.

6.2.3 A Special Case: $\mu = 1.0$

Finally, consider the situation when mutation is always operating on every allele. Then Eq. 6.1 becomes:

$$E_s[B_k, \mu = 1.0] \;=\; \sum_{q=0}^{k} P(Q = q)[\; 0^k \;+\; \left(\frac{1}{C-1}\right)^q \; 0^{k-q}\;]$$

The expression in brackets is not zero only at $q = k$. Thus:

$$E_s[B_k, \mu = 1.0] \;=\; P(Q = k) \left(\frac{1}{C-1}\right)^k$$

Now $P(Q = k) \;=\; (1 - P_{eq})^k$. Thus:

$$E_s[B_k, \mu = 1.0] \;=\; \left(\frac{1 - P_{eq}}{C-1}\right)^k \tag{6.3}$$

Note that for $C = 2$, $E_s[B_k, \mu = 1.0] = 1.0$ when $P_{eq} = 0.0$, regardless of k. This is true because if $C = 2$ and $P_{eq} = 0.0$ the two individuals are bit-wise complements of each other (at the k defining positions). Since a mutation rate of 1.0 will flip every bit, one offspring will have to be in H_k. Clearly this is not true for $C > 2$ (again compare Fig. 6.2 with Fig. 6.3). Note also that for reasonable values of P_{eq} (i.e., $> 1/C$) this is the minimum possible survival for mutation. Finally, one can see that $E_s[B_k, \mu = 1.0]$ decreases towards 0.0 as P_{eq} ranges from 0.0 to 1.0. All of these observations are intuitively obvious.

6.3 Summary

This chapter computes the expected number of offspring $E_s[B_k, \mu]$ that reside in H_k after mutation, given that one parent is a member of H_k and that the other parent is arbitrary. As would be expected, $E_s[B_k, \mu]$ decreases as the order k of the hyperplane increases, while $E_s[B_k, \mu]$ increases as P_{eq} increases, for reasonable values of μ. Interestingly, increasing the cardinality C does decrease $E_s[B_k, \mu]$, but only to a small degree, suggesting that mutation is actually not greatly affected by changes in cardinality. The results also indicate that hyperplanes are most likely to survive (high values of $E_s[B_k, \mu]$) with low levels of mutation, while high levels of mutation are most likely to disrupt hyperplanes (low values of $E_s[B_k, \mu]$). Chapter 3 provided a similar analysis for recombination using the $E_s[B_k]$ framework. The comparison between recombination and mutation (using $E_s[B_k]$) will be done in Chap. 8.

As was pointed out in Chap. 3, a more positive role of recombination is that it may construct hyperplanes from other lower-order hyperplanes. This effect was analyzed in Chap. 4. The related computation for mutation will occur in the next chapter, Chap. 7.

7. A Construction Schema Theory for Mutation

7.1 Introduction

Chapter 4 computed the probability $P_c(H_k)$ that a kth-order hyperplane H_k will be constructed via recombination, given that one parent is a member of a lower-order hyperplane H_m and that the other parent is a member of another lower-order hyperplane H_n. That chapter concluded by computing the expected number of offspring $E_c[B_k]$ that reside in H_k after recombination, and showed that this is a simple function of $P_c(H_k)$, the population homogeneity P_{eq}, and the order of the hyperplane k.

The goal of this chapter is to provide a similar computation for mutation. As stated in Chap. 6, mutation will work on alphabets of cardinality C in the following fashion. An allele is picked for mutation with probability μ. Then that allele is changed to one of the other $C-1$ alleles, uniformly randomly.[1] Mutation is performed independently to both parents, since in almost all EAs mutation is applied independently to every individual in the population.

We omit the intermediary step of computing $P_c(H_k)$ for mutation, because (as explained in Chap. 6) $P_c(H_k)$ obscures the comparison between mutation and recombination. Instead, in this chapter we will compute $E_c[B_k]$ directly.

To summarize – in order to provide for a fair comparison between recombination and mutation, mutation will be treated as if it were a two-parent operator that produces two offspring. Precisely the same random experiment will be performed as with recombination. This chapter will then use this framework to compute the expected number of offspring $E_c[B_k]$ that are in a hyperplane H_k, given that one parent is a member of a lower-order hyperplane H_m and that the other parent is a member of another lower-order hyperplane H_n, after mutation has changed the parents.

7.2 Framework

Suppose that one is given two parents, and that one parent is in the hyperplane H_m, while the other parent is in the hyperplane H_n. We will consider

[1] Again, this form of mutation is reasonable for discrete representations; however, it should be modified for real-valued representations.

the creation of a kth-order hyperplane H_k from these two hyperplanes. We will restrict the situation such that the two lower-order hyperplanes H_m and H_n are nonoverlapping, and $k = m + n$.

As stated in Chap. 4, these situations will be described by random variable $\mathcal{S}$. For a kth-order hyperplane H_k there are 2^k possible situation events: $0 \leq \mathcal{S} \leq 2^k - 1$. Each situation event $\mathcal{S}$ can be represented by a bit mask of length k. There will be m 1s and n 0s in the binary representation of $\mathcal{S}$, indicating H_m and H_n.

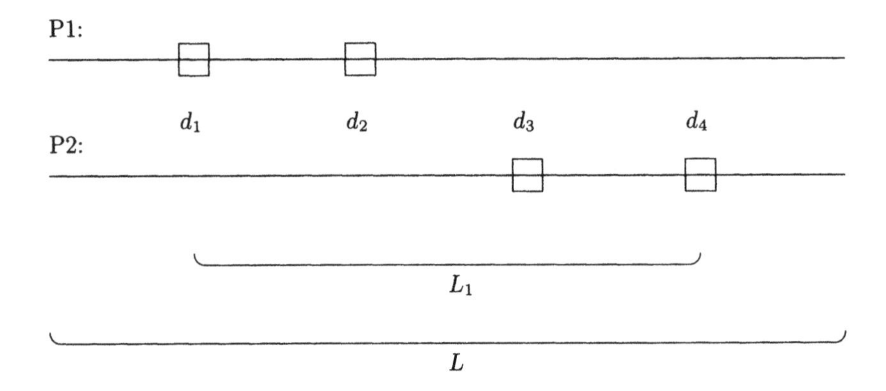

Fig. 7.1. The setup for the random experiment to be performed. P1 is a member of a second-order hyperplane, while P2 is a member of another second-order hyperplane. The goal is to construct the fourth-order hyperplane with mutation.

Figure 7.1 provides a pictorial example. The two parents are labeled P1 and P2. P1 is a member of a particular second-order hyperplane H_2, described by the alleles at defining positions d_1 and d_2. P2 is a member of another second-order hyperplane H_2, described by the alleles at defining positions d_3 and d_4. Thus this situation can be described with the binary string 1100 (or 0011), to indicate that one parent has the first two alleles, while the second parent has the second two alleles of the fourth-order hyperplane H_4.

The random experiment consists of performing mutation on these two parents, producing two children. Schema H_4 (or H_k in general) can either be created or not. A schema will be created if either offspring is in H_4 (H_k). As with recombination we will refer to this creation as a "construction." Although clearly mutation does not swap material between two parents, the term "construction" is a useful reminder that the random experiment being performed is the same as that performed in the construction analysis for recombination.

Once again, let B be a random variable describing the number of offspring that are instances of H_k. B can take on values 0, 1, and 2. We can write:

$$E[B \mid S] = \sum_{b \in \{0,1,2\}} b \times P(B = b) = P(B = 1) + 2P(B = 2)$$

Without loss of generality, assume that the first parent is in H_n, while the second parent is in H_m. For the sake of clarity, denote $E[B \mid S]$ to be $E_c[B_k, \mu \mid S]$. $E_c[B_k, \mu \mid S]$ will refer to the expected number of offspring that will be in H_k after μ mutation has been performed. The subscript 'c' is a reminder that the experiment being performed is the same as that performed for the construction analysis under recombination.

To compute $E_c[B_k, \mu \mid S]$ it will be convenient to let Q be a random variable that describes the set of alleles (at the defining positions) in the second individual that do not match H_n. Similarly, let R be a random variable that describes the set of alleles (at the defining positions) in the first individual that do not match H_m. Then we can write $E_c[B_k, \mu \mid S]$ as follows:

$$E_c[B_k, \mu \mid S] = \sum_Q \sum_R P(Q \wedge R)[\, P(B = 1 \mid Q \wedge R) + 2P(B = 2 \mid Q \wedge R)\,]$$

It will be noted that $P(Q \wedge R)$ depends on the population homogeneity. Deriving precise expressions for the values of $P(Q \wedge R)$ at a particular point in time is difficult in general since they vary from generation to generation in complex, nonlinear, and interacting ways. We can, however, get considerable insight into the effects of shared alleles on construction analysis by making simplifying assumptions.

First, as opposed to concentrating on sets, let Q be a random variable that describes the *number* of alleles (at the defining positions) in the second individual that do not match H_n. Q can take on the integer values from 0 to n. Similarly, let R be a random variable that describes the *number* of alleles (at the defining positions) in the first individual that do not match H_m. R can take on the integer values from 0 to m. For example, suppose the first parent is AABA, which is a member of the second-order hyperplane AA## (H_n), while the second parent is BAAA, which is a member of ##AA (H_m). The goal is to create an individual in AAAA. In this example Q is 1 because the second individual differs from H_n at the first defining position. Also, R is 1 because the first individual differs from H_m at the third defining position. Thus we can rewrite $E_c[B_k, \mu \mid S]$ as follows:

$$E_c[B_k, \mu \mid S] = \sum_{q=0}^{n} \sum_{r=0}^{m} P(Q = q \wedge R = r) \times$$
$$[\, P(B = 1 \mid Q = q \wedge R = r) + 2P(B = 2 \mid Q = q \wedge R = r)\,]$$

Let $P_{\text{eq}}(d)$ represent the probability that both parents have the same allele at a particular defining position d. Then further assume that $P_{\text{eq}}(d)$ is roughly the same for all the defining positions ($P_{\text{eq}}(d) = P_{\text{eq}},\ \forall\ d$).[2] Then $P(Q = q \wedge R = r)$ is simply:

$$P(Q = q \wedge R = r) \ = \ \binom{n}{q}\binom{m}{r}(1 - P_{\text{eq}})^{q+r}\ {P_{\text{eq}}}^{k-q-r}$$

Now consider the derivation of the other terms of $E_c[B_k, \mu \mid S]$. In order to have *both* offspring be in H_k (i.e., $B = 2$), the n alleles in the first parent (associated with the hyperplane H_n) must not be mutated. Also, of the remaining m alleles in the first parent, R must be mutated (while $m - R$ are not). Finally, the m alleles in the second parent (associated with the hyperplane H_m) must not be mutated. Of the remaining n alleles in the second parent, Q must be mutated (while $n - Q$ are not). To consider the example given above, the As in the two parents must not be mutated, while the Bs in each parent must be mutated to As (since one is interested in the hyperplane AAAA). For a general alphabet of cardinality C, if an allele is mutated, there is a $1/(C-1)$ probability of mutating it to the desired allele. Thus, the probability of placing *both* offspring in H_k is simply computed as:

$$P(B = 2 \mid Q = q \wedge R = r) \ =$$
$$\left[\left(\frac{\mu}{C-1}\right)^r (1-\mu)^{m-r}\,(1-\mu)^n\right]\left[\left(\frac{\mu}{C-1}\right)^q (1-\mu)^{n-q}\,(1-\mu)^m\right]$$

The first term expresses the probability of placing the first parent in H_k. The probability of not mutating the n correct alleles of the first parent is $(1-\mu)^n$. Also, since R of the remaining m alleles are incorrect, R must be mutated to the correct allele while $m - R$ are not mutated. The second term expresses the probability of placing the second parent in H_k.

It is now possible to compute the probability that *only one* offspring will be in H_k. Clearly that will occur if the first parent is placed in H_k while the second parent is not, or if the second parent is placed in H_k while the first parent is not. This can be easily computed by using the components of the previous equation:

$$P(B = 1 \mid Q = q \wedge R = r) =$$
$$\left[\left(\frac{\mu}{C-1}\right)^r (1-\mu)^{m-r}(1-\mu)^n\right]\left[1 - \left(\frac{\mu}{C-1}\right)^q (1-\mu)^{n-q}(1-\mu)^m\right] +$$
$$\left[1 - \left(\frac{\mu}{C-1}\right)^r (1-\mu)^{m-r}(1-\mu)^n\right]\left[\left(\frac{\mu}{C-1}\right)^q (1-\mu)^{n-q}(1-\mu)^m\right]$$

With some simplification $E_c[B_k, \mu \mid S]$ can now be expressed:

[2] The same assumptions were made in the recombination analysis.

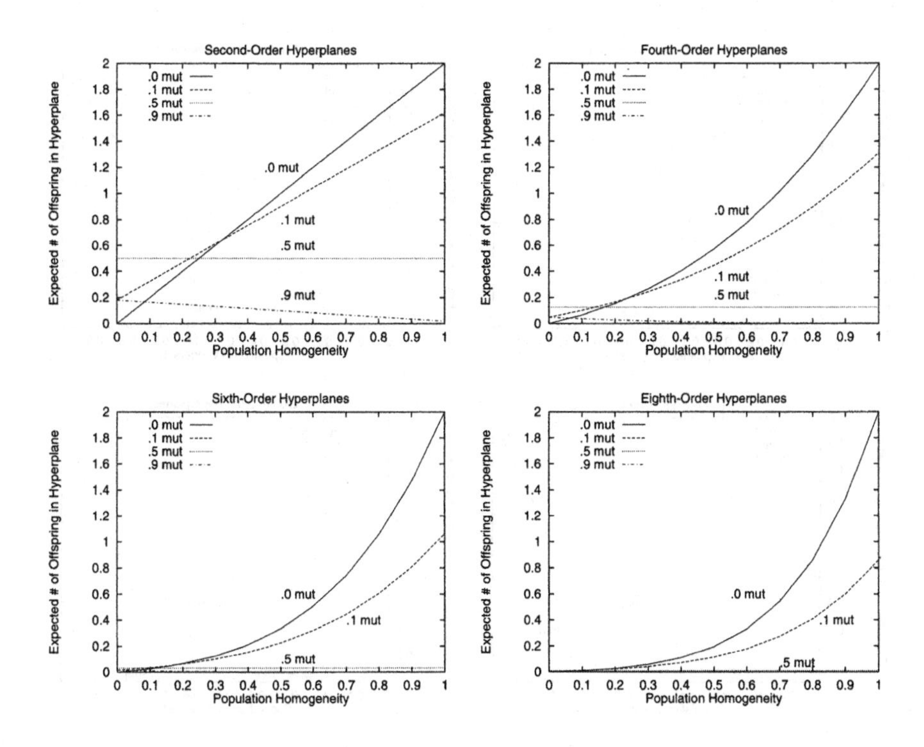

Fig. 7.2. $E_c[B_k, \mu]$ of H_2, H_4, H_6, and H_8 for mutation when $C = 2$

$$E_c[B_k, \mu \mid \mathcal{S}] = \sum_{q=0}^{n} \sum_{r=0}^{m} P(Q = q \wedge R = r) \times \left[\left(\frac{\mu}{C-1}\right)^r (1-\mu)^{k-r} + \left(\frac{\mu}{C-1}\right)^q (1-\mu)^{k-q} \right] \tag{7.1}$$

As stated in Chap. 4, of the 2^k situations $\mathcal{S}$, two of them are considered to be survival situations, and not construction situations. This occurs when $\mathcal{S} = 0$ and when $\mathcal{S} = 2^k - 1$. In the first situation $H_n = H_k$ and in the second situation $H_m = H_k$. Thus $\mathcal{S} = 0$ and $\mathcal{S} = 2^k - 1$ represent survival situations. All the other situations $(0 < \mathcal{S} < 2^k - 1)$ represent true constructions, in which part of H_k is represented by one parent, while the remainder is represented by the other parent.

Just as with the analysis of recombination in Chap. 4, it is convenient to average the results over all $2^k - 2$ construction situations. Each of the $2^k - 2$ construction situations is considered to be equally likely. This leads to:

$$E_c[B_k, \mu] = \frac{1}{2^k - 2} \sum_{\mathcal{S}=1}^{2^k-2} E_c[B_k, \mu \mid \mathcal{S}] \tag{7.2}$$

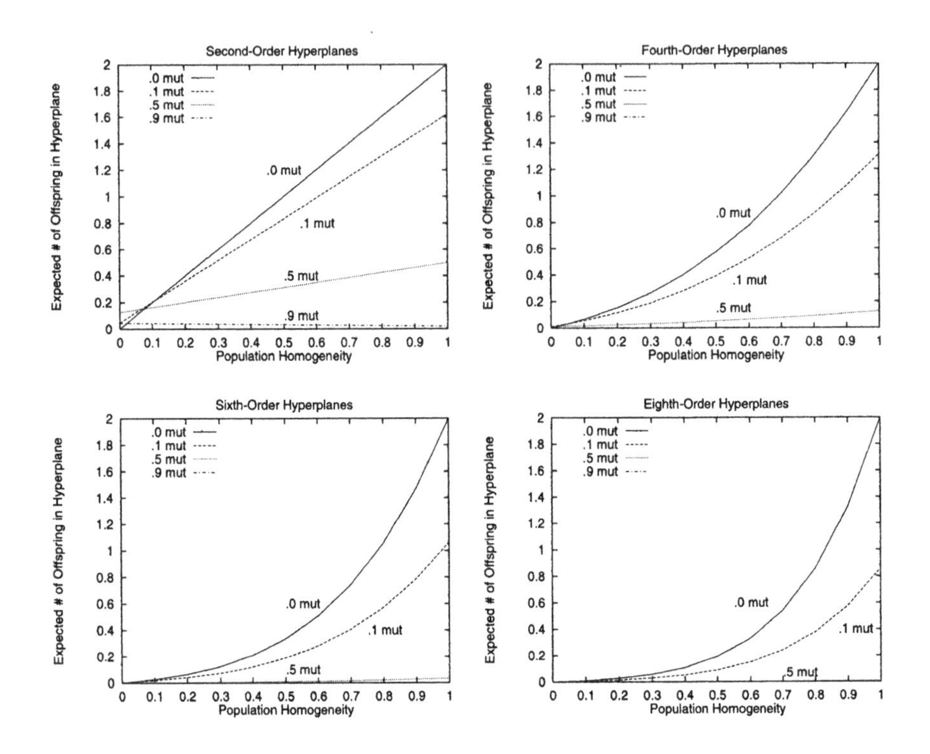

Fig. 7.3. $E_c[B_k, \mu]$ of H_2, H_4, H_6, and H_8 for mutation when $C = 5$

Figure 7.2 illustrates $E_c[B_k, \mu]$ when the cardinality of the alphabet $C = 2$, for mutation rates ranging from 0.0 to 1.0, for $k \in \{2, 4, 6, 8\}$, while P_{eq} ranges from 0.0 to 1.0. Figure 7.3 illustrates this for $C = 5$. For a randomly initialized population any allele has probability $1/C$ of being the same as any other allele, so the minimum P_{eq} is simply $1/C$. Since the populations of traditional EAs tend to become more homogeneous with time, it is reasonable to examine only those situations where $P_{eq} > 1/C$.

A number of observations can be made from the figures. The expected number of offspring in H_k ($E_c[B_k, \mu]$) decreases as the order k of the hyperplane increases. $E_c[B_k, \mu]$ increases as P_{eq} increases, for values of μ less than 0.5 (higher values are very rare in practice). Both of the observations are intuitively reasonable. More interestingly, increasing the cardinality C does decrease $E_c[B_k, \mu]$, but only to a small degree, suggesting that mutation is actually not greatly affected by changes in cardinality.

Inspection of the figures where $P_{eq} > 1/C$ indicates that the maximum construction (highest $E_c[B_k, \mu]$) occurs when $\mu = 0.0$, while the minimum construction (lowest $E_c[B_k, \mu]$) occurs when $\mu = 1.0$, as would be expected. Furthermore, the maximum construction (when $\mu = 0.0$) does not appear to depend on the cardinality of the alphabet C. This is intuitively plausible, since if no mutation is occurring, mutation will never have to choose between the $C-1$ other alleles.

Further insights can be gained by considering special cases.

7.2.1 A Special Case: $\mu = 0.0$

One nice feature of μ mutation is that it can be turned off simply by setting μ to 0.0. In this case Eq. 7.1 is:

$$E_c[B_k, \mu = 0.0 \mid \mathcal{S}] = \sum_{q=0}^{n}\sum_{r=0}^{m} P(Q = q \wedge R = r)[\left(\frac{\mu}{C-1}\right)^r + \left(\frac{\mu}{C-1}\right)^q]$$

The first term in the brackets is 1 only if $r = 0$; else it is 0. The second term in the brackets is 1 only if $q = 0$; else it is 0. Thus:

$$E_c[B_k, \mu = 0.0 \mid \mathcal{S}] = \sum_{q=0}^{n} P(Q = q \wedge R = 0) + \sum_{r=0}^{m} P(Q = 0 \wedge R = r)$$

$$E_c[B_k, \mu = 0.0 \mid \mathcal{S}] = \sum_{q=0}^{n} \binom{n}{q} (1-P_{eq})^q\, {P_{eq}}^{k-q} + \sum_{r=0}^{m} \binom{m}{r} (1-P_{eq})^r\, {P_{eq}}^{k-r}$$

This is simply:

$$E_c[B_k, \mu = 0.0 \mid \mathcal{S}] = {P_{eq}}^m \sum_{q=0}^{n} \binom{n}{q} (1-P_{eq})^q {P_{eq}}^{n-q} + {P_{eq}}^n \sum_{r=0}^{m} \binom{m}{r} (1-P_{eq})^r {P_{eq}}^{m-r}$$

$$E_c[B_k, \mu = 0.0 \mid \mathcal{S}] = {P_{eq}}^m + {P_{eq}}^n \tag{7.3}$$

This does not depend on C, but only on the population homogeneity and the order of the building blocks H_m and H_n. This is reasonable, since mutation is in fact turned off.

Now we can average these results over all construction situations:

$$E_c[B_k, \mu = 0.0] = \frac{1}{2^k - 2} \sum_{\mathcal{S}=1}^{2^k-2} E_c[B_k, \mu = 0.0 \mid \mathcal{S}]$$

Each situation $\mathcal{S}$ refers to some H_m and H_n. For mutation only the order m (or n) matters, so the above equation can be simplified:

$$E_{\mathrm{c}}[B_k, \mu = 0.0] = \frac{1}{2^k - 2} \sum_{m=1}^{k-1} \binom{k}{m} ({P_{\mathrm{eq}}}^m + {P_{\mathrm{eq}}}^{k-m})$$

$$E_{\mathrm{c}}[B_k, \mu = 0.0] =$$
$$\frac{1}{2^k - 2} \left[\sum_{m=1}^{k-1} \binom{k}{m} {P_{\mathrm{eq}}}^m 1^{k-m} + \sum_{m=1}^{k-1} \binom{k}{m} {P_{\mathrm{eq}}}^{k-m} 1^m \right]$$

Rewriting the sums to range from $m = 0$ to $m = k$ yields:

$$E_{\mathrm{c}}[B_k, \mu = 0.0] = \frac{1}{2^k - 2} \times$$
$$\left[-2 - 2{P_{\mathrm{eq}}}^k + \sum_{m=0}^{k} \binom{k}{m} {P_{\mathrm{eq}}}^m 1^{k-m} + \sum_{m=0}^{k} \binom{k}{m} {P_{\mathrm{eq}}}^{k-m} 1^m \right]$$

Using the binomial expansion yields:

$$E_{\mathrm{c}}[B_k, \mu = 0.0] = \frac{1}{2^k - 2} \left[-2 - 2{P_{\mathrm{eq}}}^k + 2(1 + P_{\mathrm{eq}})^k \right]$$

$$E_{\mathrm{c}}[B_k, \mu = 0.0] = \frac{(1 + P_{\mathrm{eq}})^k - {P_{\mathrm{eq}}}^k - 1}{2^{k-1} - 1} \tag{7.4}$$

For reasonable levels of population homogeneity ($P_{\mathrm{eq}} > 1/C$) this yields the best construction for mutation (see Fig. 7.2 and Fig. 7.3). Note how the value approaches 2.0 as P_{eq} approaches 1.0.

One other interesting simplification occurs if $k = 2$. In that case:

$$E_{\mathrm{c}}[B_2, \mu = 0.0] = (1 + P_{\mathrm{eq}})^2 - {P_{\mathrm{eq}}}^2 - 1 = 2\, P_{\mathrm{eq}}$$

which is a linear function. This can be observed in Figs. 7.2 and 7.3.

7.2.2 A Special Case: $\mu = 0.5$

Again, considerable simplifications to Eq. 7.1 can be made in the case where $\mu = 0.5$:

$$E_{\mathrm{c}}[B_k, \mu = 0.5 \mid S] =$$
$$\sum_{q=0}^{n} \sum_{r=0}^{m} P(Q = q \wedge R = r) [\frac{0.5^k}{(C-1)^r} + \frac{0.5^k}{(C-1)^q}]$$

$$E_c[B_k, \mu = 0.5 \mid \mathcal{S}] =$$
$$0.5^k \sum_{q=0}^{n} \sum_{r=0}^{m} P(Q = q \wedge R = r)[\left(\frac{1}{C-1}\right)^r + \left(\frac{1}{C-1}\right)^q]$$

In the case that $C = 2$ then:

$$E_c[B_k, \mu = 0.5 \mid \mathcal{S}] = 0.5^k + 0.5^k = \frac{1}{2^{k-1}}$$

Note how this does not depend on P_{eq}. This makes sense, since with a mutation rate μ of 0.5 and $C = 2$ the offspring are always randomly reinitialized, no matter what the parents are. This is not true when $C > 2$.

7.2.3 A Special Case: $\mu = 1.0$

When mutation is always on, Eq. 7.1 becomes:

$$E_c[B_k, \mu = 1.0 \mid \mathcal{S}] =$$
$$\sum_{q=0}^{n} \sum_{r=0}^{m} P(Q = q \wedge R = r)[\frac{0^{k-r}}{(C-1)^r} + \frac{0^{k-q}}{(C-1)^q}]$$

The first term in the brackets is 1 only if $r = k$; else it is 0. The second term in the brackets is 1 only if $q = k$; else it is 0. However, this can only occur in survival situations, and not during construction. Thus:

$$E_c[B_k, \mu = 1.0 \mid \mathcal{S}] = 0.0$$

Averaged over all construction situations:

$$E_c[B_k, \mu = 1.0] = 0.0 \tag{7.5}$$

This makes sense, since if every allele is mutated, it is impossible to have all k alleles (of H_k) on either offspring. Thus a mutation rate of 1.0 represents the worst possible construction.

7.3 Summary

This chapter computes the expected number of offspring $E_c[B_k, \mu]$ that reside in H_k after mutation, given that one parent is a member of a lower-order hyperplane H_m and that the other parent is a member of another lower-order hyperplane H_n. This is referred to as "construction," since the framework is the same as that investigated for construction of hyperplanes via recombination in Chap. 4.

The results indicate that $E_c[B_k, \mu]$ decreases as the order k of the hyperplane increases, while $E_c[B_k, \mu]$ increases as P_{eq} increases, for reasonable values of μ. Interestingly, increasing the cardinality C does decrease $E_c[B_k, \mu]$, but only to a small degree, suggesting that mutation is actually not greatly affected by changes in cardinality.

Chapter 4 found that more disruptive recombination operators achieve higher levels of construction. However, this is not the case for mutation. Although high levels of mutation are the most disruptive (low values of $E_s[B_k, \mu]$), they also achieve the worst levels of construction (lowest values of $E_c[B_k, \mu]$). This points out an interesting difference between recombination and mutation. Turning these operators off results in the lowest levels of disruption (i.e., no disruption at all). Turning on recombination increases disruption while increasing construction. However, turning on mutation increases disruption while decreasing construction.

These results indicate that, unlike recombination (see Chap. 5), a No Free Lunch theorem with respect to the disruptive and constructive aspects of mutation probably does not hold. We investigate this issue further in Chap. 8, which provides a thorough comparison of recombination and mutation via schema theory.

8. Schema Theory: Mutation versus Recombination

8.1 Introduction

The previous chapters have fully outlined schema theories for recombination and for mutation. Two aspects of these operators were investigated – the disruptive aspect and the constructive aspect. Disruption refers to the likelihood that a kth-order hyperplane H_k will not survive either recombination or mutation. Construction refers to the likelihood that a kth-order hyperplane will be created, given that one parent is a member of a lower-order hyperplane H_m and that the other parent is a member of another lower-order hyperplane H_n. In order to provide for a fair comparison of mutation and recombination, care was taken to ensure that the mathematical framework was always the same for both operators (e.g., both operators are taken to be two-parent operators that produce two children). The framework is not repeated here (see Chaps. 3–7 for full details).

Given the groundwork in those chapters, it is now possible to compare mutation and recombination via those schema theories, with respect to both the disruptive and constructive aspects of those operators. We will compare mutation with P_0 uniform recombination (as opposed to n-point recombination) because uniform recombination is easier to deal with mathematically and graphically, due to its lack of dependence on the defining lengths of hyperplanes. Since P_0 uniform recombination generally bounds the behavior of n-point recombination, this focus is quite reasonable.

8.2 Survival

Survival analysis involves computing the expected number of offspring $E_s[B_k]$ that are in a hyperplane H_k, given that one parent is in H_k and the other parent is arbitrary, after either recombination or mutation has changed the parents.

Mutation will work on alphabets of cardinality C in the following fashion. An allele is picked for mutation with probability μ. Then that allele is changed to one of the other $C - 1$ alleles, uniformly randomly. For mutation the expected number of offspring in H_k was shown in Eq. 6.1:

$$E_s[B_k, \mu] = \sum_{q=0}^{k} P(Q = q)[\ (1-\mu)^k + \left(\frac{\mu}{C-1}\right)^q (1-\mu)^{k-q}\ ] \quad (8.1)$$

For an explanation of this equation, see Chap. 6. The first parent is a member of the hyperplane H_k, while the second parent is arbitrary. Q is a random variable that describes the number of alleles (at the defining positions) in the second parent that do not match H_k. Q can take on the integer values from 0 to k. The probability distribution of Q is given by the binomial distribution:

$$P(Q = q) = \binom{k}{q} (1 - P_{\text{eq}})^q \, {P_{\text{eq}}}^{k-q}$$

where $P(Q = q)$ depends on the population homogeneity P_{eq}, which is the probability that two parents will match alleles at each defining position.

As pointed out in Chap. 6, $E_s[B_k, \mu]$ (survival) is maximized (disruption is minimized) when mutation is turned off ($\mu = 0.0$). For that situation Eq. 6.2 showed:

$$E_s[B_k, \mu = 0.0] = 1 + {P_{\text{eq}}}^k \quad (8.2)$$

In comparison, Chap. 3 showed that maximum survival (minimum disruption) occurs when P_0 uniform recombination is turned off ($P_0 = 0.0$). For that situation Eq. 3.15 and Eq. 3.17 showed:

$$E_s[B_k, P_0 = 0.0] = 1 + {P_{\text{eq}}}^k \quad (8.3)$$

This is as would be expected, since not using mutation should be equivalent to not using recombination. Thus the best rates of survival can be achieved by not using either mutation or recombination.

On the other hand, raising μ or P_0 increases the disruption caused by those operators. The maximum disruption (minimal survival) for mutation occurs when $\mu = 1.0$ (see Chap. 6), where Eq. 6.3 showed:

$$E_s[B_k, \mu = 1.0] = \left(\frac{1 - P_{\text{eq}}}{C-1}\right)^k \quad (8.4)$$

The maximum disruption for P_0 uniform recombination occurs when $P_0 = 0.5$. Higher rates of P_0 are equivalent to a rate of $1 - P_0$, and do not lead to increased disruption (see Chap. 3). In this situation Eq. 3.16 and Eq. 3.17 showed:

$$E_s[B_k, P_0 = 0.5] = \frac{2(1 + P_{\text{eq}})^k}{2^k} \quad (8.5)$$

Figure 8.1 provides a comparison of mutation and P_0 uniform recombination on $E_s[B_k]$ as the population homogeneity increases (P_{eq} increases to

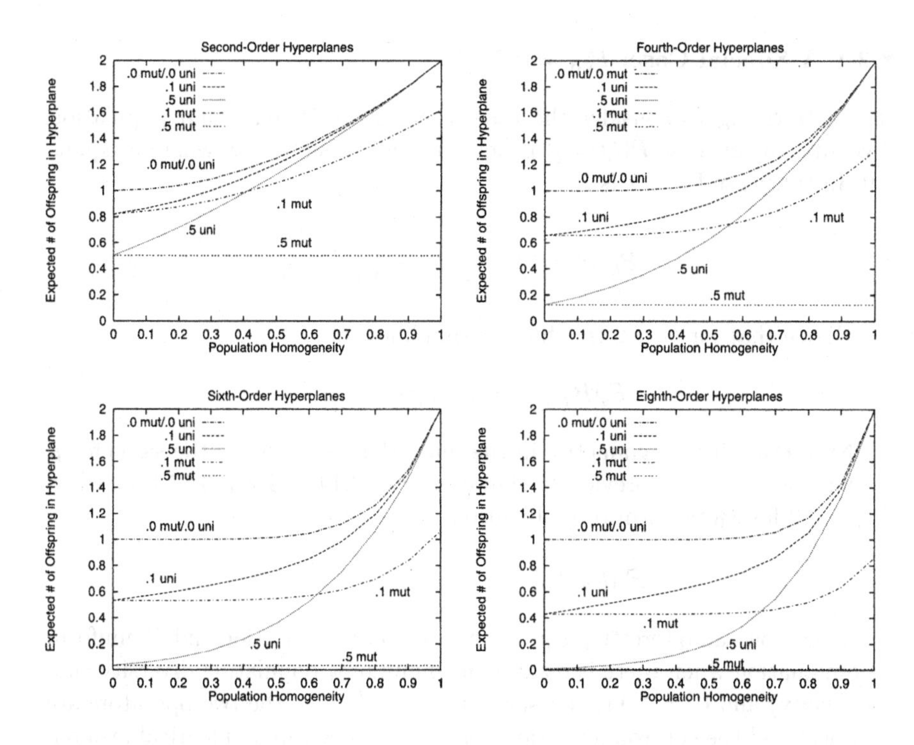

Fig. 8.1. $E_s[B_k]$ of H_2, H_4, H_6, and H_8 for mutation and recombination when $C = 2$

1.0). A comparison is made on hyperplanes of order $k \in \{2, 4, 6, 8\}$ and when the cardinality $C = 2$. For each graph three different settings of P_0 and μ are shown. As expected, the curves are identical when $P_0 = \mu = 0$. However, for any setting of P_0, uniform recombination approaches $E_s[B_k] = 2$ as the population converges. This makes sense, since recombination can only swap alleles, and when the population has converged all individuals (strings) are very similar to each other. However, this isn't true for mutation. In fact, as seen in Eq. 8.4, with $\mu = 1.0$, $E_s[B_k, \mu = 1.0]$ decreases towards 0.0 as P_{eq} ranges from 0.0 to 1.0.

Thus, in summary, one can see from this comparison that P_0 uniform recombination (as well as n-point recombination) actually has a limited range of disruptive behavior. Mutation can achieve the same maximum levels of survival (minimum disruption) as recombination can achieve. However, mutation can also achieve much lower rates of survival (higher rates of disruption) than can recombination. Thus, from a survival point of view, recombination does not appear to have any added value. The next subsection will investigate whether this is also true of the constructive aspects of recombination and mutation. Before that, however, it is instructive to investigate a particular special case of the two operators.

8.2.1 A Special Case: $P_{eq} = 0.0$

It is interesting to consider the case where there is maximum population diversity. In this case $P(Q = q)$ is not zero only when $q = k$, where it equals 1.0. In that case Eq. 8.1 is:

$$E_s[B_k, \mu] = (1 - \mu)^k + \left(\frac{\mu}{C-1}\right)^k$$

Notice that for $C = 2$ this further simplifies to:

$$E_s[B_k, \mu] = (1 - \mu)^k + \mu^k$$

Note that this is symmetric in the sense that the value is the same for μ as it is for $1 - \mu$. We previously showed in Eq. 3.14 and Eq. 3.17 that when $P_{eq} = 0.0$ for P_0 uniform recombination:

$$E_s[B_k, P_0] = (1 - P_0)^k + {P_0}^k$$

This shows an interesting equivalence between μ mutation and P_0 uniform recombination when there is maximum (minimum) population diversity (homogeneity) and $C = 2$. From a survival point of view, the two operators are identical and the controlling parameters μ and P_0 act in an identical fashion. The next section will investigate whether this also holds for construction.

8.3 Construction

Construction analysis involves computing the expected number of offspring $E_c[B_k]$ that are in a hyperplane H_k, given that one parent is a member of a lower-order hyperplane H_m and that the other parent is a member of another lower-order hyperplane H_n, after either recombination or mutation has changed the parents. The two lower-order hyperplanes are assumed to be nonoverlapping and it is assumed that $k = m + n$.

The two lower-order hyperplanes are described by a situation $\mathcal{S}$ $(0 \leq \mathcal{S} \leq 2^k - 1)$, in which the binary representation of $\mathcal{S}$ represents which parent has the necessary alleles at the k defining positions. There will be m 1s and n 0s in the binary representation of $\mathcal{S}$, indicating H_m and H_n.

There are 2^k situations, since each situation can be described via a binary string of length k. As stated in Chap. 4, of the 2^k situations $\mathcal{S}$, two of them are considered to be survival situations, and not construction situations. This occurs when $\mathcal{S} = 0$ and when $\mathcal{S} = 2^k - 1$. In the first situation $H_n = H_k$ and in the second situation $H_m = H_k$. Thus $\mathcal{S} = 0$ and $\mathcal{S} = 2^k - 1$ represent survival situations. All the other situations $(0 < \mathcal{S} < 2^k - 1)$ represent true constructions, in which part of H_k is represented by one parent, while the remainder is represented by the other parent.

The average expected number of offspring in H_k, $E_c[B_k]$, is computed by averaging over the $2^k - 2$ constructive situations $\mathcal{S}$:

$$E_c[B_k] = \frac{1}{2^k - 2} \sum_{\mathcal{S}=1}^{2^k-2} E_c[B_k \mid \mathcal{S}]$$

For mutation, the expected number of offspring in H_k (given a situation $\mathcal{S}$) is given by Eq. 7.1:

$$E_c[B_k, \mu \mid \mathcal{S}] = \sum_{q=0}^{n} \sum_{r=0}^{m} P(Q = q \wedge R = r) \times \left[\left(\frac{\mu}{C-1}\right)^r (1-\mu)^{k-r} + \left(\frac{\mu}{C-1}\right)^q (1-\mu)^{k-q} \right] \quad (8.6)$$

For an explanation of this equation, see Chap. 7. The first parent is a member of H_n, while the second parent is a member of H_m. Q is a random variable that describes the number of alleles (at the defining positions) in the second individual that do not match H_n. Q can take on the integer values from 0 to n. R is a random variable that describes the number of alleles (at the defining positions) in the first individual that do not match H_m. R can take on the integer values from 0 to m (see Chap. 7 for further details). The probability distribution of Q and R is given by:

$$P(Q = q \wedge R = r) = \binom{n}{q}\binom{m}{r} (1 - P_{\text{eq}})^{q+r} \, {P_{\text{eq}}}^{k-q-r}$$

where $P(Q = q \wedge R = r)$ depends on the population homogeneity P_{eq}, which is the probability that two parents will match alleles at each defining position.

As pointed out in Chap. 7, construction is minimized when mutation is always on ($\mu = 1.0$). In that case Eq. 7.5 says:

$$E_c[B_k, \mu = 1.0] = 0.0 \quad (8.7)$$

The best construction for mutation (for reasonable levels of $P_{\text{eq}} > 1/C$) occurs when $\mu = 0.0$ (see Chap. 7). Equation 7.3 and Eq. 7.4 say:

$$E_c[B_k, \mu = 0.0 \mid \mathcal{S}] = {P_{\text{eq}}}^m + {P_{\text{eq}}}^n \quad (8.8)$$

$$E_c[B_k, \mu = 0.0] = \frac{(1 + P_{\text{eq}})^k - {P_{\text{eq}}}^k - 1}{2^{k-1} - 1} \quad (8.9)$$

It is interesting to contrast this with the *worst* construction for P_0 uniform recombination, which occurs at $P_0 = 0.0$. As shown in Eq. 4.7 and Eq. 4.12 of Chap. 4:

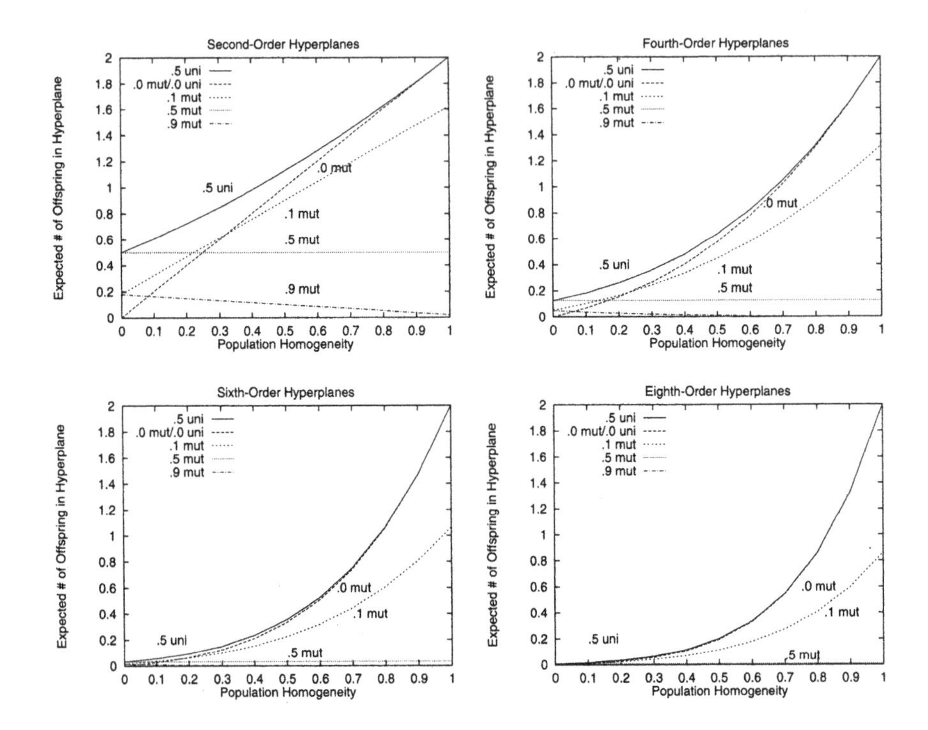

Fig. 8.2. $E_c[B_k]$ of H_2, H_4, H_6, and H_8 for mutation and recombination when $C = 2$

$$E_c[B_k, P_0 = 0.0 \mid \mathcal{S}] \;=\; P_{\text{eq}}{}^n \;+\; P_{\text{eq}}{}^m \tag{8.10}$$

Note that since this is exactly the same as when $\mu = 0.0$ (as would be expected) we can conclude that for $P_0 = 0.0$ uniform recombination:

$$E_c[B_k, P_0 = 0.0] \;=\; \frac{(1 + P_{\text{eq}})^k \;-\; P_{\text{eq}}{}^k \;-\; 1}{2^{k-1} \;-\; 1} \tag{8.11}$$

This is an important observation. What we have shown is that the *worst* construction for P_0 uniform recombination is precisely the same as the *best* construction for μ mutation!

Now, as shown in Chap. 4, construction increases as P_0 is increased from 0.0 to 0.5. The best construction for P_0 uniform recombination occurs at $P_0 = 0.5$ (see Eq. 4.11 and Eq. 4.13):

$$E_c[B_k, P_0 = 0.5] \;=\; \frac{2(1 + P_{\text{eq}})^k}{2^k} \tag{8.12}$$

What we have shown is that recombination is clearly better at construction than mutation. The worst construction for recombination matches the best construction for mutation. The constructive advantage of recombination

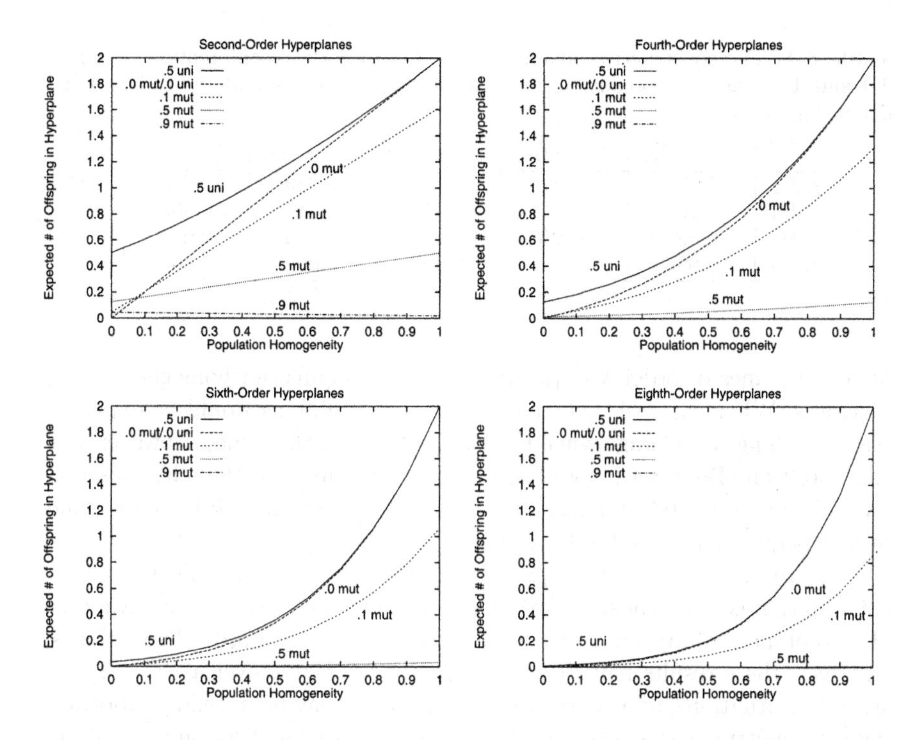

Fig. 8.3. $E_c[B_k]$ of H_2, H_4, H_6, and H_8 for mutation and recombination when $C = 5$

can be seen in Fig. 8.2 and Fig. 8.3. The figures graph $E_c[B_k]$ for different settings of μ mutation and P_0 uniform recombination on hyperplanes of order $k \in \{2, 4, 6, 8\}$. Alphabets of cardinality $C = 2$ are assumed in Fig. 8.2, whereas a cardinality of $C = 5$ is shown in Fig. 8.3. The cardinality of the alphabet will not affect the results for recombination, but it will affect the mutation results. In general the highest two curves in a graph represent $E_c[B_k]$ for 0.0 mutation and 0.5 uniform recombination (the best rates for both operators). The highest curve is always that for 0.5 uniform recombination. Again, the focus is always on that portion of the graphs which reflects reasonable levels of population homogeneity (i.e., $P_{eq} > 1/C$).

It is now possible to derive an expression yielding the constructive advantage Γ of the most constructive recombination operator (0.5 uniform recombination) versus the most constructive mutation operator (0.0 mutation):

$$\Gamma = \frac{2(1 + P_{eq})^k}{2^k} - \frac{(1 + P_{eq})^k - {P_{eq}}^k - 1}{2^{k-1} - 1} \tag{8.13}$$

Note that this depends only on the population homogeneity and the order of the hyperplane H_k. It does not depend on the cardinality of the alphabet, so this can be ignored. The results for Eq. 8.13 are shown in Table 8.1

Table 8.1. The constructive advantage Γ of 0.5 uniform recombination over $\mu = 0.0$ mutation, as the population homogeneity P_{eq} increases and the order of the hyperplane k increases

	$P_{\mathrm{eq}} = 0.2$	$P_{\mathrm{eq}} = 0.4$	$P_{\mathrm{eq}} = 0.6$	$P_{\mathrm{eq}} = 0.8$	$P_{\mathrm{eq}} = 1.0$
H_2	0.320	0.180	0.080	0.020	0.000
H_4	0.106	0.078	0.044	0.014	0.000
H_6	0.029	0.025	0.017	0.006	0.000
H_8	0.008	0.007	0.005	0.002	0.000

for hyperplanes of order $k \in \{2, 4, 6, 8\}$, as the population homogeneity P_{eq} increases. When $P_{\mathrm{eq}} = 1.0$ the advantage goes to 0.0, as would be expected (since nothing will change when the two parents are the same). However, one can note from Table 8.1 (as well as from Figs. 8.2 and 8.3) that the constructive advantage for 0.5 uniform recombination decreases as k increases, and in fact is quite small by the time that $k = 8$.

This result is somewhat surprising. Although the results indicate that recombination is more constructive than mutation, the constructive advantage of recombination appears to be small, especially for high-order hyperplanes. However, these results have been obtained by averaging over all $2^k - 2$ construction situations, and the averaging procedure may be masking important details concerning the relative advantages of recombination and mutation. Thus it is worthwhile to ask how that advantage is distributed over those situations. This is explored in the next subsection.

8.3.1 The Distribution Over All Situations

As shown earlier, the expected number of offspring in H_k produced with mutation (given a situation $\mathcal{S}$) is given by Eq. 8.6:

$$E_{\mathrm{c}}[B_k, \mu \mid \mathcal{S}] = \sum_{q=0}^{n} \sum_{r=0}^{m} P(Q = q \wedge R = r) \times \left[\left(\frac{\mu}{C-1}\right)^{r} (1-\mu)^{k-r} + \left(\frac{\mu}{C-1}\right)^{q} (1-\mu)^{k-q} \right]$$

This can easily be compared with uniform recombination on a situation by situation basis. As shown earlier in Eq. 4.5 of Chap. 4:

$$P_{\mathrm{c}}(H_k, P_0 \mid \mathcal{S}) = P_{\mathrm{s,orig}}(H_m, P_0)\, P_{\mathrm{s,other}}(H_n, P_0) + P_{\mathrm{s,other}}(H_m, P_0)\, P_{\mathrm{s,orig}}(H_n, P_0) - {P_{\mathrm{eq}}}^{k}$$

where:

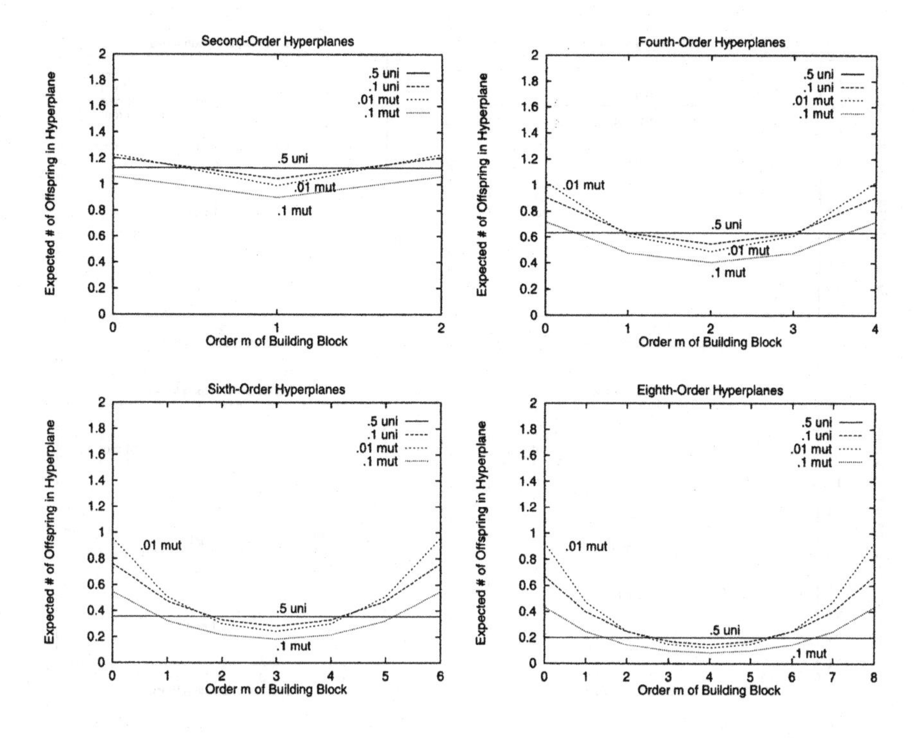

Fig. 8.4. $E_{c,s}[B_k \mid \mathcal{S}]$ of H_2, H_4, H_6, and H_8 for mutation and recombination as m varies, when $C = 2$ and $P_{eq} = 0.5$

$$P_{s,orig}(H_k, P_0) = \sum_{x=0}^{k} \binom{k}{x} P_0^{\,x}(1 - P_0)^{k-x} P_{eq}^{\;x}$$

$$P_{s,other}(H_k, P_0) = \sum_{x=0}^{k} \binom{k}{x} P_0^{\,x}(1 - P_0)^{k-x} P_{eq}^{\;k-x}$$

Thus the expected number of offspring in H_k for uniform recombination is (by using Eq. 4.12):

$$E_c[B_k, P_0 \mid \mathcal{S}] = P_{s,orig}(H_m, P_0)\, P_{s,other}(H_n, P_0) + P_{s,other}(H_m, P_0)\, P_{s,orig}(H_n, P_0)$$

Now $E_c[B_k \mid \mathcal{S}]$ for mutation is not affected by the defining lengths of hyperplanes. $E_c[B_k \mid \mathcal{S}]$ for uniform recombination is also not affected by the defining lengths. However, both are affected by the order of hyperplanes. This makes a comparison between the two operators over all situations very straightforward.

Figure 8.4 plots the expected number of offspring in H_k produced by mutation and P_0 uniform recombination when $C = 2$ and $P_{eq} = 0.5$, as the order m of the hyperplane H_m ranges from 0 to k (and $n = k - m$). To be

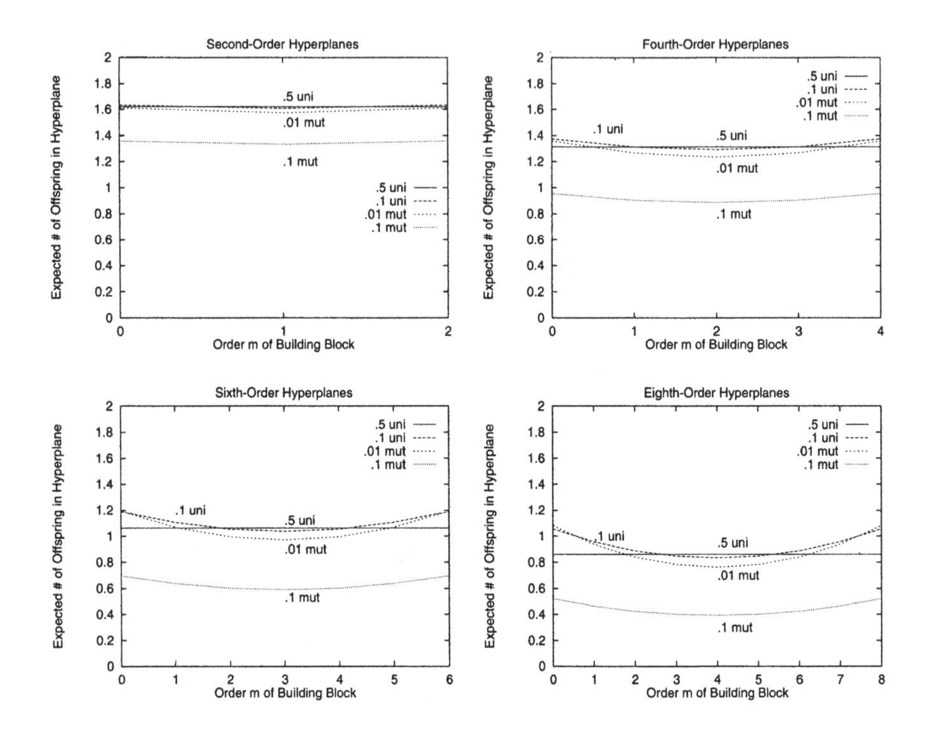

Fig. 8.5. $E_{c,s}[B_k \mid S]$ of H_2, H_4, H_6, and H_8 for mutation and recombination as m varies, when $C = 2$ and $P_{eq} = 0.8$

complete, we deliberately include those situations that are actually not construction situations but are survival situations (when $m = 0$ or $m = k$). Thus the graph allows for a comparison of mutation and uniform recombination on all possible situations.

What is interesting about this figure is that it nicely illustrates the situations where recombination most dramatically outperforms mutation from a constructive perspective, namely, when the order of the building blocks is about 1/2 of k (i.e., $n = m$). As the two building blocks H_m and H_n become more unequal in order, mutation starts to become more advantageous! This is quite reasonable, since if one building block has a very low order, constructive effects can occur with a minimal number of mutations. However, recombination is much more likely to combine two building blocks of relatively high order.

Figure 8.5 illustrates roughly the same behavior when $P_{eq} = 0.8$. However, one observation is that the region wherein recombination is advantageous both broadens and flattens, providing less advantage, but over more situations. As P_0 is reduced from 0.5 to 0.1 it trades off construction for survival, i.e., it is less constructive than 0.5 uniform recombination in the middle of the graph, but shows better survival qualities at the extremes. This is a

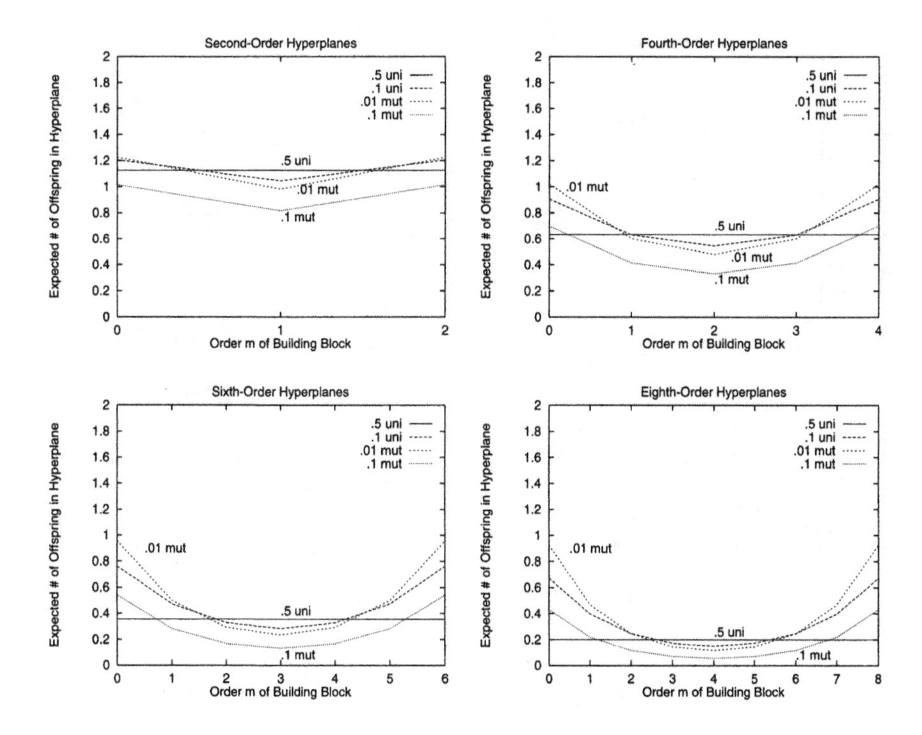

Fig. 8.6. $E_{\mathrm{c,s}}[B_k \mid \mathcal{S}]$ of H_2, H_4, H_6, and H_8 for mutation and recombination as m varies, when $C = 20$ and $P_{\mathrm{eq}} = 0.5$

nice demonstration of the No Free Lunch theorem for recombination that was presented in Chap. 5. In contrast, Fig. 8.5 indicates that $E_{\mathrm{c,s}}[B_k \mid \mathcal{S}]$ for $\mu = 0.1$ mutation is lower than $\mu = 0.01$ mutation for every situation $\mathcal{S}$. Thus, in general, mutation does not obey a similar No Free Lunch theorem (construction is not traded off against survival).

To see how the cardinality of the alphabet affects matters (it will affect mutation only), we also tried $C = 20$. As shown in Chap. 7, the constructive aspects of mutation should decrease as C increases, since it becomes increasingly more difficult to mutate an undesirable allele to a desirable allele. Figure 8.6 and Fig. 8.7 illustrate this for a P_{eq} of 0.5 and 0.8. Although the hypothesis is confirmed, it is surprising how well mutation holds up – the decrease in constructive ability is very small.

8.3.2 A Special Case: $P_{\mathrm{eq}} = 0.0$

The previous section (concerning survival) showed an interesting equivalence between μ mutation and P_0 uniform recombination when there is maximum population diversity ($P_{\mathrm{eq}} = 0.0$) and minimal cardinality ($C = 2$). From a survival point of view, the two operators are identical and the controlling

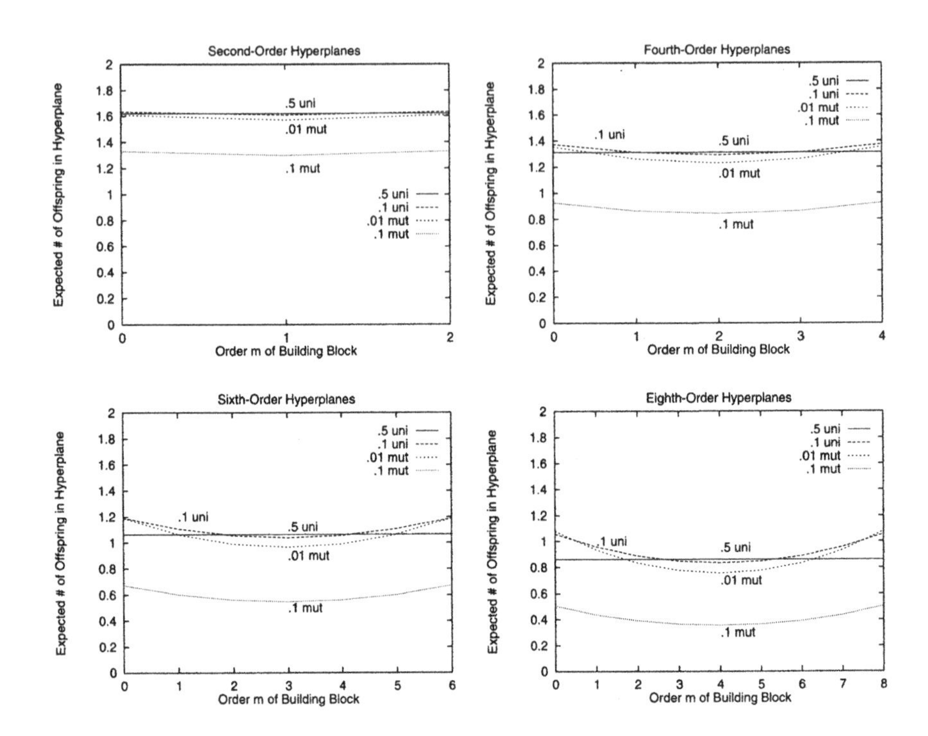

Fig. 8.7. $E_{c,s}[B_k \mid \mathcal{S}]$ of H_2, H_4, H_6, and H_8 for mutation and recombination as m varies, when $C = 20$ and $P_{eq} = 0.8$

parameters μ and P_0 act in an identical fashion. This subsection investigates whether this is also true for the constructive aspects of recombination and mutation.

Consider when there is maximum population diversity: $P_{eq} = 0.0$. In this case $P(Q = q \wedge R = r)$ is not zero only when $q + r = k$ (in which case it equals 1.0). This means that $r = m$ and $q = n$. In that case Eq. 8.6 is:

$$E_c[B_k, \mu \mid \mathcal{S}] = \left(\frac{\mu}{C-1}\right)^m (1-\mu)^{k-m} + \left(\frac{\mu}{C-1}\right)^n (1-\mu)^{k-n}$$

When $C = 2$:

$$E_c[B_k, \mu \mid \mathcal{S}] = \mu^m (1-\mu)^{k-m} + \mu^n (1-\mu)^{k-n}$$

$$E_c[B_k, \mu \mid \mathcal{S}] = \mu^m (1-\mu)^n + \mu^n (1-\mu)^m$$

However, Eq. 4.6 and Eq. 4.12 of Chap. 4 previously showed that when $P_{eq} = 0.0$, $E_c[B_k, P_0 \mid \mathcal{S}]$ for P_0 uniform recombination is:

$$E_c[B_k, P_0 \mid \mathcal{S}] = {P_0}^m (1-P_0)^n + {P_0}^n (1-P_0)^m$$

Once again we see an interesting equivalence between the constructive aspect of μ mutation and P_0 uniform recombination when there is maximum population diversity and $C = 2$. The two parameters (μ and P_0) act the same way and produce identical results.

8.4 Survival and Construction

It was previously shown in Chap. 5 that recombination obeys a No Free Lunch theorem, in the sense that any increase in survival (decrease in disruption) is offset by a decrease in construction. This was shown by computing the average of $E_{c,s}[B_k \mid S]$ over all situations $\mathcal{S}$ (including survival and construction). Equation 5.6 showed that, for all forms of recombination:

$$E_{c,s}[B_k] = \frac{2(1+P_{eq})^k}{2^k}$$

Thus, for any form of recombination, $E_{c,s}[B_k]$ is simply a constant that depends on the order of the hyperplane H_k and the population homogeneity. Since the average is taken over both survival and construction situations, this indicates that any increase in survival must be offset by a commensurate decrease in construction. As shown in the last section, however, this is *not* true for mutation, in general. However, there is one special case in which mutation does obey a No Free Lunch theorem. The previous two sections showed the equivalence of mutation and P_0 uniform recombination when $P_{eq} = 0.0$ and $C = 2$, from both a survival and constructive aspect. Thus it would be expected that mutation would also obey a No Free Lunch theorem when $P_{eq} = 0.0$ and $C = 2$. This in fact is true. As shown in the last section, when $P_{eq} = 0.0$ and $C = 2$, Eq. 8.6 is:

$$E_{c,s}[B_k, \mu \mid S] = \mu^m (1-\mu)^{k-m} + \mu^n (1-\mu)^{k-n}$$

Thus, averaging over all situations $\mathcal{S}$ yields:

$$E_{c,s}[B_k, \mu] = \frac{1}{2^k} \sum_{m=0}^{k} \binom{k}{m} [\, \mu^m (1-\mu)^{k-m} + \mu^n (1-\mu)^{k-n} \,]$$

This is trivially simplied to:

$$E_{c,s}[B_k, \mu] = \frac{2}{2^k} = \frac{1}{2^{k-1}}$$

Thus, no matter what the value of μ, the average expected number of offspring in H_k after mutation is again a constant that only depends on the order of the hyperplane H_k. Note that as would be expected, this is the same expression that is obtained for recombination when $P_{eq} = 0.0$ (see Eq. 5.7).

8.5 Summary

This chapter has provided a full comparison of mutation and recombination via the schema theories developed in the previous chapters. Both the disruptive and constructive aspects of mutation and recombination were compared.

The results indicate that, from a survival point of view, recombination is not as powerful as mutation. Mutation can achieve any level of survival (disruption) that recombination can achieve. Moreover, mutation can achieve higher levels of disruption than recombination. However, from a construction point of view, recombination is more powerful than mutation. In fact the worst levels of construction for recombination are the same as the best levels of construction for mutation. This is achieved when both operators are turned off ($P_0 = 0.0$ and $\mu = 0.0$). Increasing P_0 from 0.0 increases construction until a maximum is reached at $P_0 = 0.5$. However, increasing μ from 0.0 decreases construction. Construction becomes impossible when $\mu = 1.0$.

Simply put, recombination is able to recombine lower-order building blocks into higher-order building blocks with higher expectation than can mutation. When averaged over all possible constructive situations this advantage is often quite small. An analysis of the distribution of construction over all situations provides a more useful picture – the largest constructive advantage for recombination occurs when both lower-order building blocks have order roughly 1/2 of k (the order of H_k).

This paints the picture that recombination will be most useful when high-fitness building blocks of relatively high order (H_m and H_n) can be combined into higher-order building blocks (H_k) that are also of high fitness. Recombination will be least useful when the higher-order building blocks that are constructed have poor fitness. It is important to point out that the "fitness" of a hyperplane refers to its "observed" fitness (i.e., the average fitness of those individuals within a population that are in that hyperplane) as opposed to its "static" fitness (i.e., the average fitness of *all* individuals within that hyperplane). We will see how this observation helps us to create problems that are both easy and hard for recombination (see Chap. 12 and Chap. 14).

Another interesting observation arising from this analysis is that μ mutation and P_0 uniform recombination are identical (from both a survival and constructive aspect) when binary strings are used ($C = 2$) and there is maximum population diversity ($P_{\mathrm{eq}} = 0.0$). Interestingly, similar results will also be shown with respect to other static analyses of P_0 uniform recombination and μ mutation in Chap. 9. However, it is important to realize that this maximum level of population diversity is somewhat unusual – it assumes that the two parents are different at every allele from each other. This is unlikely to occur with standard EAs. However, speciating EAs are much more likely to maintain individuals at radically different areas of the search space. Thus, it is quite possible that this observation will be more useful for speciating EAs (e.g., see Spears 1994).

The key to the schema analyses that have been provided thus far has been the computation of the expected number of offspring that will be in H_k after recombination or mutation. It is this computation that allows for a fair comparison between mutation and recombination. However, the comparison has been between two parents undergoing only mutation and two parents undergoing only recombination. The joint behavior of the two operators acting in concert has not been investigated thus far. Fortunately, it appears as if the framework developed in this book could be extended, by computing the expected number of offspring in H_k after applying both recombination *and* mutation to two parents. This will not be explored in this book, but will be deferred for future work.

It should be noted that the schema analyses provided in this chapter (and in the prior chapters) can be considered to be *static* analyses, since they do not take into account the dynamic time evolution of an EA. This book will also investigate *dynamic* analyses (Chaps. 10–12). However, there are still other static analyses to consider. These are investigated in Chap. 9.

9. Other Static Characterizations of Mutation and Recombination

9.1 Introduction

The previous chapters have provided a full schema analysis of recombination and mutation. Such analyses can be considered to be static in the sense that they do not take into account the dynamic time evolution of an EA. However, there have been other static characterizations of recombination that have been explored in the literature (such as "exploratory power," "positional bias," and "distributional bias"). The point of this chapter is to extend these characterizations of recombination to mutation, in order to compare the two operators.

Several of the prior analyses have made two assumptions for n-point recombination that are a bit different from those used in this book. This book has assumed (see Chap. 3) that there are L possible distinct cut-points for n-point recombination. The Lth cut-point can be considered to occur immediately before the individual or immediately after the individual. Booker (1992) and Eshelman et al. (1989) both appear to assume that there are only $L-1$ possible cut-points (the Lth cut-point is considered impossible). The second assumption is that we choose cut-points with replacement, whereas some analyses (Booker 1992) assume sampling without replacement.

One consequence of our assumptions is that certain n-point recombination events will not change the parents. For example, this can occur in two-point recombination if the same cut-point is chosen twice. However, this is reasonable, since P_0 uniform recombination also may not change the parents (if none or all of the alleles are swapped). Another consequence is that any recombination event that is possible under n-point recombination is also possible under $(n+1)$-point recombination. For example, consider the two parents: AABB and ABAB. The one-point recombination event AA|BB and AB|AB is the same as the two-point recombination event AA|BB| and AB|AB|. Thus, two-point recombination can achieve any effect that one-point recombination can achieve (but with lower probability).

Generally, the first assumption does not lead to qualitative differences in results. However, the second assumption can lead to differences, especially as n approaches L for n-point recombination. In the limit where $n = L$, there is only one possible recombination event if the cut-points are chosen without replacement. If cut-points are chosen with replacement, there are an enormous number of possible recombination events. However, for most applications of EAs, $n \ll L$, and the differences between sampling with and without replacement are negligible. Thus, to be consistent with the remainder of the book we will continue to assume that there are L possible cut-points and that the n cut-points are sampled with replacement.

9.2 Exploratory Power

Perhaps one of the simplest characterizations of an EA operator is its "exploratory power," which is defined to be the number of different individuals that can potentially be created by one application of that operator. Eshelman et al. (1989) investigated the exploratory power of recombination. This section carefully re-examines this work and then extends the characterization to include mutation.

9.2.1 The Exploratory Power of Recombination

The "exploratory power" of recombination is defined to be the number of different individuals that can be created by one application of recombination to two parents. The immediate observation is that the population homogeneity is crucial. If there is no diversity, then the two parents will be the same, and recombination cannot create any new individuals. As population homogeneity decreases, more and more individuals can be reached through recombination.

As pointed out in the earlier chapters, a useful measure of population homogeneity is denoted as $P_{\text{eq}}(d)$, which represents the probability that both parents have the same allele at a particular defining position d. For the sake of analysis it is also useful to further assume that $P_{\text{eq}}(d)$ is roughly the same for all the defining positions ($P_{\text{eq}}(d) = P_{\text{eq}}, \forall\ d$). Thus $P_{\text{eq}} = 0.0$ represents maximum diversity (minimum homogeneity), while $P_{\text{eq}} = 1.0$ represents minimum diversity (maximum homogeneity).

Let Y be a random variable representing the number of alleles that are different in the two parents (which is equivalent to Hamming distance when the cardinality is $C = 2$). Y can range from 0 (if both parents are identical) to L (if none of the alleles match between the two parents). The probability distribution of Y depends on P_{eq}:

$$P(Y = y) = \binom{L}{y} {P_{\text{eq}}}^{L-y}(1 - P_{\text{eq}})^y$$

Clearly, if there are y different alleles, any recombination operator can produce at most 2^y individuals, since recombination can only swap alleles. For example, suppose the language has cardinality $C = 4$ with an alphabet of {A,B,C,D}. Further suppose that two parents AAAAA and AABCD are recombined. Then $y = 3$ and the maximum number of potentially reachable strings is 2^3. The exploratory power of any recombination operator increases as the population diversity (and hence y) increases. The maximum exploratory power occurs if $y = L$, which occurs when $P_{\text{eq}} = 0.0$. The cardinality of the language has no effect on this result.

The exploratory power of a particular recombination operator is related to the number of recombination events that are possible with that recombination operator (see Chap. 3 for a definition of a recombination event). The more recombination events that are possible, the higher the exploratory power of that operator. To see this, consider one-point recombination. There are L possible recombination events (because there are L possible cut-points). If y of the L alleles are different (between the two parents), then $y - 1$ of those events can produce new individuals (the yth event doesn't change the parents). To continue the above example, only two recombination events can produce new individuals. The first event is when the cut-point is after the third allele: AAA|AA and AAB|CD. The second event is when the cut-point is after the fourth allele: AAAA|A and AABC|D. Since there are two parents, one application of one-point recombination can reach at most $2(y - 1)$ new individuals. In this example, where $y = 3$, four new individuals can be created with one-point recombination: AAACD, AAAAD, AABAA, and AABCA.

Similarly, there are $L(L+1)/2$ possible recombination events for two-point recombination (because we are sampling cut-points with replacement). Only $y(y+1)/2$ of those events will be relevant if only y of the L alleles are different. Also, y of those events will produce no effect (when the same cut-point location is chosen for both cut-points), so only $y(y+1)/2 - y = y(y-1)/2$ recombination events can produce new children. Since there are two parents, one application of two-point recombination can reach at most $y(y - 1)$ new individuals. To continue the above example, where $y = 3$, only three recombination events can produce new children. The first event is AAA|AA| and AAB|CD|. The second event is AAAA|A| and AABC|D|. These two events are the same as the two events for one-point recombination. The third event is AAA|A|A and AAB|C|D. Thus, six new individuals can be created with two-point recombination: AAACA, AABAD, and the individuals produced from one-point recombination.

We have shown that one-point recombination has very low exploratory power. The exploratory power of two-point recombination is somewhat higher. This is as expected, since we know that the set of possible recombination events for two-point recombination is a superset of the set of possible events for one-point recombination. Thus, in general, since the set of possible events under $(n+1)$-point recombination always includes the events from n-point recombination, the exploratory power of n-point recombination increases as n increases.[1]

[1] This would not be true if we sample cut-points without replacement, since only one event would be possible for L-point recombination in that situation.

As we have shown earlier, if y of the L alleles differ between the two parents, the maximum exploratory power possible for any recombination operator is 2^y. In general, n-point recombination has an exploratory power less than this maximum, although it does increase as n increases. P_0 uniform recombination, however, is quite different. When P_0 is 0.0 or 1.0, uniform recombination has no exploratory power, since it cannot change the parents. However, for $0.0 < P_0 < 1.0$, there are 2^L possible recombination events (even if some events are low probability, they are nonzero probability). If y alleles differ between the two parents, P_0 uniform recombination can reach 2^y individuals (when P_0 isn't 0.0 or 1.0). For example, if two parents are AAAAAA and AABCDE, then $y = 4$, and 2^4 individuals can be reached. Thus P_0 uniform recombination can achieve the maximum possible exploratory power. In general, P_0 uniform recombination is far more explorative than n-point recombination (unless n is very high).

9.2.2 The Exploratory Power of Mutation

Mutation will work on alphabets of cardinality C in the following fashion. An allele is picked for mutation with probability μ. Then that allele is changed to one of the other $C - 1$ alleles, uniformly randomly.

The exploratory power of mutation is simple to compute. If the mutation rate μ is 0.0, no alleles will change, and there is no exploration. If the mutation rate is 1.0 then all alleles will change, and $(C-1)^L$ individuals can be reached (by applying mutation to one parent). However, if $0.0 < \mu < 1.0$ the number of individuals that can be reached is C^L (since alleles may or may not change). In other words, any individual is potentially reachable via mutation from any other individual. These results are clearly not affected by the population homogeneity, although they are dramatically affected by the cardinality of the alphabet.

Now it is possible to compare the exploratory power of mutation and recombination. In general, mutation has much higher exploratory power than recombination, since it is possible for any individual to be created with the application of mutation to one individual. This is especially true when the population starts to converge. In that case recombination can produce almost no new individuals, while mutation still can. It is also more dramatic for high-cardinality alphabets, since recombination can produce at most 2^L individuals, while mutation can produce at most C^L individuals.

There is one situation, however, where recombination and mutation have the same exploratory power. Earlier in this book, Chap. 8 showed that mutation and P_0 uniform recombination perform in the same way (from a schema point of view) if the population has maximum diversity and the cardinality $C = 2$. Interestingly, the same observation can be made again in this section. If there is maximum diversity, $P_{\text{eq}} = 0.0$, and the two parents for recombination differ at all $y = L$ alleles. Thus P_0 uniform recombination (for

$0.0 < P_0 < 1.0$) has an exploratory power of 2^L in that situation, which is the same as the exploratory power of mutation ($0.0 < \mu < 1.0$).

9.3 Positional Bias

Another characterization of an EA operator is its "positional bias," which refers to the extent that the creation of any new schema by the operator is dependent upon the location of the alleles in the chromosomes. Eshelman et al. (1989) and Booker (1992) investigated the positional bias of recombination. This section summarizes those results and then compares them with mutation.

9.3.1 The Positional Bias of Recombination

A recombination operator has positional bias to the extent that the creation of any new schema by recombining existing schemata is dependent upon the location of the alleles in the chromosomes. Booker (1992) showed that, of the n-point recombination operators, one-point recombination has the highest positional bias. Also, Booker showed that for $n < L/2$ the positional bias of n-point recombination tends to decrease as n increases.[2]

Uniform recombination is quite different from n-point recombination. Since alleles are swapped independently, and without regard to their location on the chromosomes, it is readily apparent that P_0 uniform recombination has no positional bias at all. None of the n-point recombination operators have zero positional bias. The interested reader is urged to consult Booker (1992) for more details on the positional bias of recombination operators.

Positional bias is similar to the "length" bias introduced in Chap. 3. For example, one-point recombination is far more likely to disrupt the schema A###A than the schema AA###, since the latter schema has a shorter defining length. Simply stated, the flatness (horizontalness) of the survival curves presented in Chap. 3 are a good qualitative indication of the positional bias of a recombination operator. For example, consider Fig. 3.9 which shows the survival curves for n-point recombination. The x-axis represents defining length. It is clear that as n increases, the survival curves of n-point recombination become flatter and are hence less affected by the defining length of the schemata. However, they are never totally flat, and there is always some degree of length bias.

On the other hand, Fig. 3.9 also shows the survival curves for P_0 uniform recombination. These survival curves are always flat, indicating their total lack of length bias.

[2] Booker (1992) also showed that the positional bias increased as n increases past $L/2$. This may be due to the fact that cut-points are sampled without replacement.

9.3.2 The Positional Bias of Mutation

The positional bias of mutation is quite straightforward to compute – there is none! Since the mutation of an allele is not affected by the position of the allele, mutation has no positional bias. Thus, once again μ mutation is more similar to P_0 uniform recombination than n-point recombination.

9.4 Distributional Bias

Finally, the last characterization of an operator is its "distributional bias," which is defined by examining the distribution of the alleles that are exchanged by the operator. If the distribution is uniform, there is no bias. The more the distribution differs from the uniform distribution, the higher the distributional bias. Eshelman et al. (1989) and Booker (1992) carefully investigated the distributional bias of recombination. Note, however, that simply *exchanging* alleles between two parents via recombination may not actually *change* them, especially when the population homogeneity is high (since the two parents will be very similar). Thus, only a subset of the exchanged alleles will actually change. For example, although the one-point recombination of AAA|ABC and AAA|ADE exchanges three alleles, only two alleles change in each parent. Also, mutation does not exchange alleles between two parents, but simply changes alleles. Thus, a proper framework for comparison will have to focus on the material (alleles) actually changed by either operator. This section extends the prior work by computing the distribution of the amount of material actually changed by recombination. This distribution is affected by the population homogeneity. The section concludes by performing a similar computation for mutation, allowing for a comparison between the two operators.

9.4.1 The Distributional Bias of Recombination

A recombination operator has distributional bias to the extent that the amount of material that is expected to be exchanged is distributed around some value or values as opposed to uniformly distributed ranging from 0 to $L - 1$ alleles (where the chromosome is composed of L genes). However, in order to compare with mutation, it will be necessary to consider not only the amount of material exchanged, but the actual subset of material changed, which is a function of the population homogeneity. Suppose X is a random variable representing the amount of material *exchanged*, while Y is a random variable representing the amount of material actually *changed*. X ranges from 0 to L, while Y ranges from 0 to X. The expected amount of material Y that will be changed, given the amount of material X that is exchanged is:

$$E[Y \mid X = x] = \sum_{y=0}^{x} y \, P(Y = y \mid X = x)$$

Now the trick is to estimate the probability that y alleles will change (in each parent), given that x alleles have been exchanged. This can be accomplished using the definition of P_{eq} given earlier, which gives the probability that two alleles (one in each parent at some defining position) will be the same:

$$P(Y = y \mid X = x) = \binom{x}{y} {P_{\text{eq}}}^{x-y}(1 - P_{\text{eq}})^y$$

For y of the x exchanged alleles to actually change, $x - y$ of the alleles will have to match in both parents. The combinatorial takes into account the number of different ways to get y alleles from the x. The expected amount of material Y that will be changed, given the amount of material X that is exchanged is:

$$E[Y \mid X = x] = \sum_{y=0}^{x} y \binom{x}{y} {P_{\text{eq}}}^{x-y}(1 - P_{\text{eq}})^y \tag{9.1}$$

Finally, one can compute the expected amount of material that will be changed (as opposed to exchanged):

$$E[Y] = \sum_{x=0}^{L} E[Y \mid X = x]\; P(X = x) \tag{9.2}$$

The goal now is to compute the probability that x alleles will be exchanged, $P(X = x)$, for the various recombination operators. Again, the reader is urged to consult Booker (1992), which provides the computation of this quantity for n-point and P_0 uniform recombination. However, since our assumptions (concerning L cut-points which are sampled with replacement in n-point recombination) are a bit different from those used by Booker, we provide our own computations for one-point, two-point, and P_0 uniform recombination.

Using our assumptions of L possible cut-points that are sampled with replacement, it is easy to show that for one-point recombination:

$$P(X = x) = \frac{1}{L} \quad \forall\, x < L$$

Thus, since the distribution is uniform, there is no distributional bias for one-point recombination. Surprisingly, two-point recombination also does not have distributional bias, since it can be shown that once again:

$$P(X = x) = \frac{1}{L} \quad \forall\, x < L$$

Since both of these results agree with those found by Booker (1992), we omit the computations for higher n. Booker found that the distributional bias of n-point recombination tends to increase as n increases, as the distribution becomes less and less uniform.

The point of this framework is that with Eq. 9.2 and $P(X = x)$, one can compute the expected number of alleles actually changed via recombination, not simply the number of alleles exchanged. Thus, this framework includes the effects of population homogeneity, unlike the earlier work by Eshelman et al. (1989) and Booker (1992).

Since the previous sections have shown interesting relationships between P_0 uniform recombination and μ mutation, it is instructive to now provide the computation of $P(X = x)$ for P_0 uniform recombination:

$$P(X = x) = \binom{L}{x} P_0{}^x (1 - P_0)^{L-x}$$

Since this is a binomial distribution, P_0 uniform recombination has high distributional bias. The distributional bias increases as P_0 decreases from 0.5 to 0.0 (Booker 1992). Thus, for P_0 uniform recombination, the expected number of alleles that are changed is:

$$E[Y, P_0] = \tag{9.3}$$
$$\sum_{x=0}^{L} \sum_{y=0}^{x} y \binom{x}{y} P_{\mathrm{eq}}{}^{x-y} (1 - P_{\mathrm{eq}})^y \binom{L}{x} P_0{}^x (1 - P_0)^{L-x}$$

Thus the expected value of the number of alleles changed depends on the length of the string L, the probability of swapping alleles P_0, and the homogeneity of the population (represented using P_{eq}). It is important to note that these formulas (whether for n-point or P_0 uniform recombination) hold for arbitrary finite-cardinality alphabets.[3]

9.4.2 The Distributional Bias of Mutation

Consider now the distributional bias of mutation, which changes alleles with probability μ. One can immediately compute the probability that y alleles will be changed, which is governed by the binomial distribution:

$$P(Y = y) = \binom{L}{y} \mu^y (1 - \mu)^{L-y}$$

As an aside, if L is large ($\gg 1$) and μ is small ($\ll 1$) then the Poisson distribution can be used as an estimate for the binomial distribution, so the probability that y alleles will be changed by mutation can be approximated by:

[3] The only effect that the cardinality C has on the analysis is in determining what values of P_{eq} are relevant.

$$P(Y = y) = \frac{(L\mu)^y e^{(-L\mu)}}{y!}$$

where the expectation and variance of Y is simply $L\mu$.

However, this section will continue to use the more accurate binomial distribution to compute the expected number of alleles that will be changed:

$$E[Y, \mu] = \sum_{y=0}^{L} y\ P(Y = y) = \sum_{y=0}^{L} y \binom{L}{y} \mu^y (1 - \mu)^{L-y} \tag{9.4}$$

The expected amount of material changed depends on the string length L and the mutation rate μ. It does not depend on the population homogeneity. The result is independent of the cardinality of the alphabet, since all that matters is that an allele is chosen for mutation (and once it is chosen, it must change).

Now a comparison of mutation and P_0 uniform recombination is in order. Consider once again the special situation in which a population is maximally diverse ($P_{eq} = 0.0$). This means that all alleles that are exchanged will in fact be changed, and Eq. 9.1 is:

$$E[Y, P_0 \mid X = x] = \sum_{y=0}^{x} y \binom{x}{y} P_{eq}{}^{x-y}(1 - P_{eq})^y = x$$

Thus Eq. 9.3 becomes:

$$E[Y, P_0] = \sum_{x=0}^{L} x \binom{L}{x} P_0{}^x (1 - P_0)^{L-x} \tag{9.5}$$

One can see the immediate similarity between $E[Y, \mu]$ (Eq. 9.4), and $E[Y, P_0]$ (Eq. 9.5). The two expressions are identical in form, and P_0 serves the same purpose for P_0 uniform recombination as μ does for mutation. Thus, when parents are maximally different ($P_{eq} = 0.0$), mutation and parameterized uniform recombination change the same number of alleles (in expectation) when $P_0 = \mu$. Now, P_{eq} will be close to 0.0 when a population of individuals is randomly initialized and the alphabet is of high cardinality. In the case of binary strings, P_{eq} will initially be 0.5. As the population converges, P_{eq} will increase, thus reducing the number of alleles changed via recombination. In order to mimic the distributional bias of recombination with mutation, μ would have to slowly decrease as the population converges.

9.5 Summary

This chapter has explored other static characterizations of recombination operators, namely their "exploratory power," their "positional bias," and their "distributional bias." These characterizations indicate substantial differences between P_0 uniform recombination and n-point recombination. P_0 uniform recombination has no positional bias, while n-point recombination does. On the other hand, n-point recombination (for small n) has very low distributional bias, while P_0 uniform recombination has high distributional bias. Finally, P_0 uniform recombination tends to be far more explorative than n-point recombination – unless n is large (close to L). This confirms the earlier results of Eshelman et al. (1989) and Booker (1992).

This chapter also extended these characterizations to include mutation, allowing for a comparison of the two operators. As shown earlier in Chap. 8, μ mutation and P_0 uniform recombination are identical (in terms of the schema characterization considered earlier in this book) when binary strings are used ($C = 2$) and there is maximum population diversity ($P_{\text{eq}} = 0.0$). Interestingly, similar relationships are seen again with all three characterizations considered in this chapter. Neither mutation nor P_0 uniform recombination have positional bias. Mutation has an exploratory power of 2^L when the cardinality $C = 2$. Similarly, P_0 uniform recombination also has exploratory power 2^L when $C = 2$ and there is maximum population diversity. Finally, both mutation and P_0 uniform recombination have the same distributional bias when there is maximum population diversity and $P_0 = \mu$. In all cases mutation appears to be more similar in behavior to P_0 uniform recombination than n-point recombination.

The key contribution of this chapter is to provide a framework that extends the earlier analyses of Eshelman et al. (1989) and Booker (1992). This new framework includes the effects of population homogeneity and extends the prior analyses to cover mutation. Clearly much more work can be done with this framework – for example, explicit computations of the exploratory power of n-point recombination (where $n > 2$) can be derived. Also, it would be possible to investigate the behavior of recombination and mutation in combination. This work will be deferred until the future.

This concludes our treatment of static analyses. The remainder of the book will consider dynamic analyses, in which the time evolution of a population of chromosomes is considered. Chapter 10 considers the time evolution of populations undergoing mutation and/or recombination, without selection. Chapter 11 considers the expected time evolution of a population undergoing selection and mutation. Finally, Chapter 12 discusses a Markov framework of an evolutionary algorithm with selection, mutation, and recombination. As the reader will see, some of the lessons learned from the static analyses will serve as inspiration for the experiments and analyses performed in Chaps. 10–12, culminating in predictions concerning the performance of actual EAs (Chap. 14).

Part III

Dynamic Theoretical Analyses

10. Dynamic Analyses of Mutation and Recombination

10.1 Introduction

The previous chapters in this book have examined static characterizations of recombination and mutation. They were static in the sense that the time evolution of a population of chromosomes was not considered. Although static characterizations can be extremely helpful, it is also useful to investigate dynamic characterizations in which the time evolution is explicitly considered.

This chapter will investigate how a population of chromosomes evolves under recombination and mutation. This investigation involves examining the limiting distribution of the population as it is recombined or mutated (i.e., the expected population in the limit of infinite time). When possible we also attempt to determine which operators approach the limiting distribution more quickly than others. In doing so we find that the prior static schema analyses (see Chaps. 3 and 4) are intimately related to these dynamic analyses.

10.2 The Limiting Distribution for Recombination

Geiringer's Theorem (Geiringer 1944) describes the equilibrium distribution of an arbitrary population that is undergoing recombination, but no selection or mutation. To understand Geiringer's Theorem, consider a population of ten binary strings of length four. In the initial population, five of the strings are AAAA while the other five are BBBB. If these strings are recombined repeatedly, eventually every possible string (16 possibilities) will become equally likely in the population. In equilibrium, the probability of a particular string will approach the product of the initial probabilities of the individual alleles – thus asserting a condition of independence between alleles.

Let S be any string of L alleles: $(a_1, ..., a_L)$. Geiringer's Theorem states that if a population is recombined repeatedly (without selection or mutation) then:

$$\lim_{t \to \infty} p_S{}^{(t)} = \prod_{i=1}^{L} p_{a_i}{}^{(0)} \qquad (10.1)$$

where ${p_S}^{(t)}$ is the expected proportion of string S in the population at time t and ${p_{a_i}}^{(0)}$ is the proportion of allele a at locus (position) i in the initial population. Thus, the probability of string S is simply the product of the proportions of the individual alleles in the initial ($t = 0$) population. The equilibrium distribution illustrated in Eq. 10.1 is referred to as "Robbins' equilibrium" (Robbins 1918). Equation 10.1 holds for all standard recombination operators (such as n-point recombination and P_0 uniform recombination). It also holds for arbitrary cardinality alphabets. The key point is that recombination operators do not change the distribution of alleles at any locus, they merely shuffle those alleles at each locus.

10.2.1 The Rate at Which Robbins' Equilibrium Is Approached

Booker (1992) states that the rate at which the population approaches Robbins' equilibrium is the significant distinguishing characterization of different recombination operators. However, Booker does not attempt to actually perform that characterization. One reasonable hypothesis is that those recombination operators that are more disruptive would drive the population to equilibrium more quickly (see Christiansen 1989 and Mühlenbein 1998 for evidence to support this hypothesis). This will be investigated by examining the differential equations describing the *expected* time evolution of the strings in a population of finite size (equivalently this can be considered to be the evolution of an infinite-size population). The treatment will hold for hyperplanes as well, so the term "hyperplane" and "string" can be used interchangeably.

Consider having a population of strings. Each generation, pairs of strings (parents) are repeatedly chosen uniformly randomly for recombination, producing offspring for the next generation. Let S_h, S_i, and S_j be strings of length L (alternatively, they can be considered to be hyperplanes of order L). Let ${p_{S_i}}^{(t)}$ be the proportion of string S_i at time t. The time evolution of S_i will involve terms of loss and gain. A loss will occur if parent S_i is recombined with another parent such that neither offspring is S_i. A gain will occur if two parents that are not S_i are recombined to produce S_i. Thus the following differential equation can be written for each string S_i:

$$\frac{d{p_{S_i}}^{(t)}}{dt} = -\,{loss_{S_i}}^{(t)} + {gain_{S_i}}^{(t)} \tag{10.2}$$

The losses can occur if S_i is recombined with another string S_j such that S_i and S_j differ by $\Delta(S_i, S_j) \equiv k$ alleles, where k ranges from two to L. For example the string AAAA can (potentially) be lost if recombined with AABB (where $k = 2$). If S_i and S_j differ by one or zero alleles, there will be no change in the proportion of string S_i. In general, the expected loss for string S_i at time t is:

$$\begin{aligned} {loss_{S_i}}^{(t)} &= \\ &\sum_{S_j} {p_{S_i}}^{(t)}\, {p_{S_j}}^{(t)} P_{\mathrm{d}}(H_k) \quad \text{where } 2 \le \Delta(S_i, S_j) \equiv k \le L \end{aligned} \tag{10.3}$$

The product $p_{S_i}{}^{(t)}\ p_{S_j}{}^{(t)}$ is the probability that S_i will be recombined with S_j, and $P_{\mathrm{d}}(H_k)$ is the probability that neither offspring will be S_i. Equivalently, $P_{\mathrm{d}}(H_k)$ refers to the probability of disrupting the kth-order hyperplane H_k defined by the k different alleles. This is identical to the probability of disruption that was introduced in the static schema analysis performed in Chap. 3.

Gains can occur if two strings S_h and S_j of length L can be recombined to construct S_i. As with the earlier discussion of construction of hyperplanes (see Chap. 4), it is assumed that neither S_h or S_j is the same as S_i at all defining positions and that either S_h or S_j has the correct allele for S_i at every locus. Suppose that S_h and S_j differ at $\Delta(S_h, S_j) \equiv k$ alleles. Once again k must range from two to L. For example, the string AAAA can (potentially) be constructed from the two strings AABB and ABAA (where $k = 3$). If S_h and S_j differ by one or zero alleles, then either S_h or S_j is equivalent to S_i and there is no true construction (or gain).

Of the k differing alleles, m are at string S_h and $n = k - m$ are at string S_j. Thus what is happening is that two nonoverlapping, lower-order building blocks H_m and H_n are being constructed to form H_k (and thus the string S_i). In general, the expected gain for string S_i at time t is:

$$gains_{S_i}{}^{(t)} = \sum_{S_h, S_j} p_{S_h}{}^{(t)}\ p_{S_j}{}^{(t)}\ P_{\mathrm{c}}(H_k \mid H_m \wedge H_n) \quad \text{where } 2 \leq \Delta(S_h, S_j) \equiv k \leq L \tag{10.4}$$

The product $p_{S_h}{}^{(t)}\ p_{S_j}{}^{(t)}$ is the probability that S_h will be recombined with S_j, and $P_{\mathrm{c}}(H_k \mid H_m \wedge H_n)$ is the probability that an offspring will be S_i. Equivalently, $P_{\mathrm{c}}(H_k \mid H_m \wedge H_n)$ is the probability of constructing the kth-order hyperplane H_k (and hence string S_i) from the two strings S_h and S_j that contain the nonoverlapping, lower-order building blocks H_m and H_n. This is identical to the probability of construction that was introduced in the static schema analysis performed in Chap. 4.

If the cardinality of the alphabet is C then there are C^L different strings. This results in a system of C^L simultaneous first-order differential equations. What is important to note is the explicit connection between Eqs. 10.3–10.4 and the schema theory for recombination presented earlier. Both the probability of disruption (Chap. 3) and the probability of construction (Chap. 4) of schemata appear in the differential equations, indicating a tight link between this dynamic theory and the earlier (static) schema theory.

In general, $P_{\mathrm{d}}(H_k)$ and $P_{\mathrm{c}}(H_k \mid H_m \wedge H_n)$ depend on the population homogeneity. Earlier chapters used P_{eq} to denote the population homogeneity – P_{eq} is the probability that the two alleles match on two parents, at a locus. In this particular analysis, the strings must differ at the k alleles, so $P_{\mathrm{eq}} = 0.0$ for all k alleles (see Chaps. 3 and 4 for details).

Naturally, a closed-form solution to the system of differential equations could yield direct insights into the time evolution of the system (e.g., the rate at which Robbins' equilibrium is approached). Unfortunately, the differential equations are nonlinear, creating enormous difficulties in achieving a closed-form solution. Thus it appears to be very difficult to actually determine the exact rate at which a particular recombination operator will drive a string (or hyperplane) to Robbins' equilibrium. However, it would be of interest to answer a slightly easier question: "Will recombination operator A drive a string (hyperplane) to equilibrium faster than recombination operator B?"

Given the explicit connection that has been made between the dynamic analysis and the prior static schema analyses, it is tempting to hypothesize that the earlier graphs of survival (disruption) and construction of schemata (when $P_{eq} = 0.0$) will yield valuable insights as to how fast a particular recombination operator will drive a hyperplane to Robbins' equilibrium, relative to another recombination operator. The intuitive feeling is that a more disruptive recombination operator would drive the hyperplane to equilibrium more quickly.

If some hyperplane is above the equilibrium proportion then the loss terms will be more important, as they drive the hyperplane down to equilibrium. A more disruptive recombination operator will increase $P_d(H_k)$ and hence drive the hyperplane down towards equilibrium more quickly. Likewise, if some hyperplane is below the equilibrium proportion then the gain terms will be more important, as they drive the hyperplane up towards equilibrium. A more disruptive recombination operator can increase $P_c(H_k \mid H_m \wedge H_n)$ and hence drive that hyperplane to equilibrium more quickly. In both cases a more disruptive recombination operator can drive the system to Robbins' equilibrium more quickly.

Although this may be impossible to prove in general, it turns out that it can be demonstrated for second-order hyperplanes H_2 under n-point and P_0 uniform recombination, and for low-order hyperplanes under P_0 uniform recombination.

10.2.2 A Special Case: P_0 Uniform Recombination

For P_0 uniform recombination the loss and gain terms are especially easy to compute. As stated earlier, losses can occur if an Lth-order hyperplane S_i is recombined with an Lth-order hyperplane S_j such that S_i and S_j differ by k alleles, where k ranges from two to L. But this occurs with probability (see Eq. 3.14 of Chap. 3):

$$P_d(H_k, P_0) = 1 - P_s(H_k, P_0) = 1 - P_0{}^k - (1 - P_0)^k \qquad 2 \le k \le L$$

Once again, since the k alleles actually differ in the two hyperplanes, $P_{eq} = 0.0$ in the computation of $P_d(H_k, P_0)$ (and hence P_{eq} does not appear). Note that this expression is symmetric in the sense that $1 - P_0$ produces the same level of disruption as P_0. By taking the derivative with respect to P_0

and finding where it is 0, it can be shown that the graph of $P_d(H_k, P_0)$ has zero slope when:

$$(1 - P_0)^{k-1} = P_0{}^{k-1}$$

which occurs when $P_0 = 0.5$. Since the second derivative is less than zero everywhere, disruption is at a maximum when $P_0 = 0.5$ and disruption decreases as P_0 decreases or increases from 0.5.

Thus, the key point is that when the time evolution of the population undergoing recombination is expressed with C^L differential equations, the effect of increasing or decreasing P_0 from 0.5 reduces *all* of the loss terms in the differential equations. This slows the rate at which the equilibrium is approached.

Gains will occur if two hyperplanes S_h and S_j of order L can be recombined to construct S_i. Again, suppose that S_h and S_j differ at k alleles, where k ranges from two to L. Of the k differing alleles, m are at hyperplane S_h and $n = k - m$ are at hyperplane S_j. Then the probability of construction is (see Eq. 4.6 of Chap. 4):

$$P_c(H_k, P_0 \mid H_m \wedge H_n) = \\ P_0{}^m(1 - P_0)^n + P_0{}^n(1 - P_0)^m \quad 2 \leq k \leq L,\ 0 < m < k$$

Note that once again this expression is symmetric in the sense that $1 - P_0$ produces the same level of construction as P_0. By taking the derivative with respect to P_0 and finding where it is 0, it can be shown that $P_c(H_k, P_0 \mid H_m \wedge H_n)$ has zero slope when:

$$mP_0{}^{m-1}(1 - P_0)^n + nP_0{}^{n-1}(1 - P_0)^m = \\ nP_0{}^m(1 - P_0)^{n-1} + mP_0{}^n(1 - P_0)^{m-1}$$

which occurs when $P_0 = 0.5$.

The question now is under what conditions of n and m will $P_0 = 0.5$ represent a global (and the only) maximum. It is easy to show by counterexample that $P_0 = 0.5$ is not a global maximum for arbitrary m and n (the reader can try $m = 1$ and $n = 4$). However, there are various cases where $P_0 = 0.5$ is a global maximum – namely when $m = 1$ and $n = 1$, $m = 1$ and $n = 2$, $m = 1$ and $n = 3$, and when $m = 2$ and $n = 2$. In these cases (and the symmetric cases where m and n are interchanged) the first derivative is positive when $P_0 < 0.5$ and negative when $P_0 > 0.5$. Since we are interested in kth-order hyperplanes (where $k = m + n$), we have shown that for low-order hyperplanes ($k < 5$), construction is at a maximum when $P_0 = 0.5$ and construction decreases as P_0 decreases or increases from 0.5.

Thus, consider the time evolution of the hyperplanes in a population that are undergoing recombination, as modeled with the above differential equations. What we have shown is that if the hyperplanes have low order ($k < 5$), the effect of increasing or decreasing P_0 from 0.5 reduces *all* of

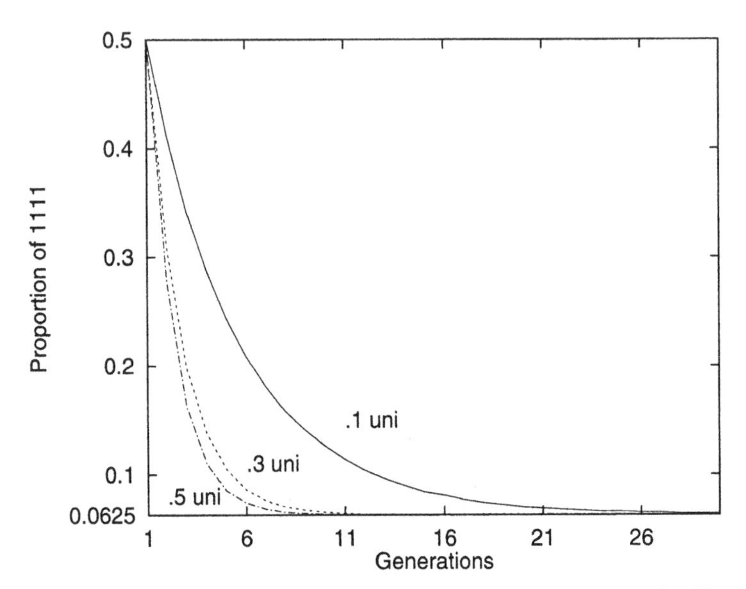

Fig. 10.1. The rate of approaching Robbins' equilibrium for H_4

the gain terms in the differential equations, slowing the rate at which the equilibrium is approached.

To summarize, for low-order hyperplanes ($k < 5$) $P_0 = 0.5$ appears to be the setting at which uniform recombination approaches equilibrium the fastest. Reducing or increasing P_0 from 0.5 will serve to decrease the rate at which the equilibrium is approached. To test this, an experiment was performed in which a population of binary strings was initialized so that 50% of the strings were all 1s, while 50% were all 0s. The strings were of length $L = 30$ and were repeatedly recombined, generation by generation, while the percentage of the fourth-order hyperplane 1111 was monitored. When Robbins' equilibrium is reached the percentage of any of the fourth-order hyperplanes should be 6.25%. The experiment was run with uniform recombination, with P_0 ranging from 0.1 to 0.5 (higher values were ignored due to symmetry).

Figure 10.1 graphs the results. One can see that as P_0 increases to 0.5, the rate at which Robbins' equilibrium is approached also increases, as expected. What we have shown, then, is that although we can't answer precisely how fast P_0 uniform recombination will approach equilibrium, we do know that decreasing (or increasing) P_0 from 0.5 will always slow the approach towards equilibrium (for $k < 5$).

It is natural to wonder how this extends to higher-order hyperplanes. Unfortunately, as pointed out above, there will be situations (of m and n) where $P_0 = 0.5$ does not represent a global maximum for construction. However, it is easy to prove that when $m = n$ the first derivative is positive when $P_0 < 0.5$ and negative when $P_0 > 0.5$, and thus once again construction decreases as P_0 decreases or increases from 0.5. It appears as if this also holds for those situations where m and n are roughly equal (i.e., both are roughly $k/2$), but eventually fails when m and n are sufficiently different (either m or n is close to 1). Since there are more situations (in a combinatorial sense) where m and n are roughly equal, there is some hope that $P_0 = 0.5$ will often still represent the fastest approach towards equilibrium for most hyperplanes when $k > 4$. However, this is still an open issue.

These results would appear to be at odds with the earlier results in Chap. 5, in which it was proven that since $P_0 = 0.5$ is the most disruptive of the uniform recombination operators, it is also the most constructive (for arbitrary order hyperplanes). However, this result was obtained by uniformly averaging construction over all possible situations (of m and n), whereas in this section we are considering each possible situation separately. Since the situations will not necessarily be uniformly distributed in the time evolution of the differential equations, the previous results cannot be applied.

10.2.3 A Special Case: Second-Order Hyperplanes

It is now natural to ask how n-point recombination compares with P_0 uniform recombination, in terms of how fast equilibrium is approached. It turns out that this can be answered easily for second-order hyperplanes.

Consider the special case where the cardinality of the alphabet $C = 2$. In this case there are four hyperplanes of interest: (#0#0#, #0#1#, #1#0#, #1#1#).[1] Then the four differential equations describing the expected time evolution of these hyperplanes are:

$$\begin{aligned}
\frac{dp_{00}{}^{(t)}}{dt} &= -p_{00}{}^{(t)}\, p_{11}{}^{(t)}\, P_{\mathrm{d}}(H_2) \;+\; p_{01}{}^{(t)}\, p_{10}{}^{(t)}\, P_{\mathrm{c}}(H_2 \mid H_1 \wedge H_1) \\
\frac{dp_{01}{}^{(t)}}{dt} &= -p_{01}{}^{(t)}\, p_{10}{}^{(t)}\, P_{\mathrm{d}}(H_2) \;+\; p_{00}{}^{(t)}\, p_{11}{}^{(t)}\, P_{\mathrm{c}}(H_2 \mid H_1 \wedge H_1) \\
\frac{dp_{10}{}^{(t)}}{dt} &= -p_{01}{}^{(t)}\, p_{10}{}^{(t)}\, P_{\mathrm{d}}(H_2) \;+\; p_{00}{}^{(t)}\, p_{11}{}^{(t)}\, P_{\mathrm{c}}(H_2 \mid H_1 \wedge H_1) \\
\frac{dp_{11}{}^{(t)}}{dt} &= -p_{00}{}^{(t)}\, p_{11}{}^{(t)}\, P_{\mathrm{d}}(H_2) \;+\; p_{01}{}^{(t)}\, p_{10}{}^{(t)}\, P_{\mathrm{c}}(H_2 \mid H_1 \wedge H_1)
\end{aligned}$$

Thus for this special case the loss and gain terms are controlled fully by one computation of disruption and one computation of construction. If two recombination operators have precisely the same disruption and construction behavior on second-order hyperplanes, the system of differential equations will be the same, and the time evolution of the system will be the same. This is true regardless of the initial conditions of the system.

[1] These four hyperplanes have been chosen arbitrarily for illustrative purposes. Also, we assume a binary-string representation, although that isn't necessary.

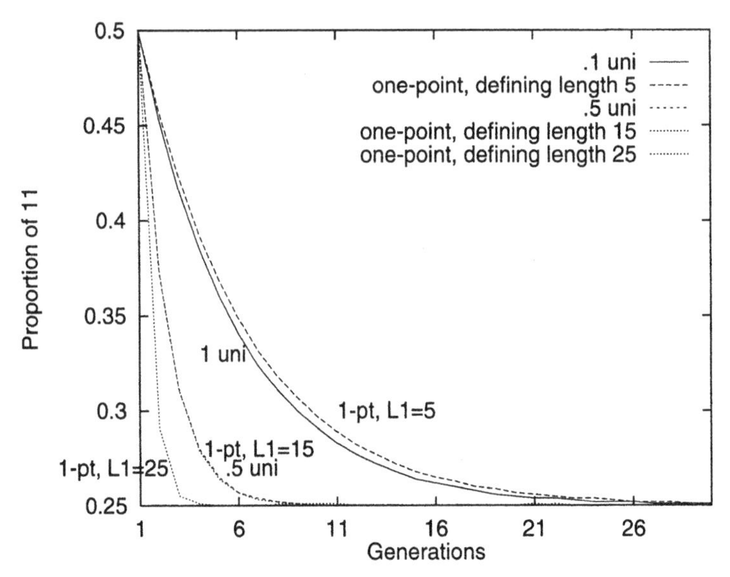

Fig. 10.2. The rate of approaching Robbins' equilibrium for H_2, when $L = 30$

For example, consider one-point recombination and P_0 uniform recombination. Suppose the defining length of the second-order hyperplane is L_1. Then, since $P_{eq} = 0.0$, $P_d(H_2) = L_1/L$ for one-point recombination, and $P_d(H_2) = 2P_0(1-P_0)$ for uniform recombination (see Eq. 3.14 when $k = 2$). The computations for $P_c(H_2 \mid H_1 \wedge H_1)$ yield identical results. Thus, one-point recombination should act the same as uniform recombination when the defining length $L_1 = 2LP_0(1 - P_0)$.

To test this, an experiment was performed in which a population of binary strings was initialized so that 50% of the strings were all 1s, while 50% were all 0s. The strings were of length $L = 30$ and were repeatedly recombined, generation by generation, while the percentage of the second-order hyperplane #1#1# was monitored. When Robbins' equilibrium is reached the percentage of any of the four hyperplanes should be 25%. The experiment was run with 0.1 and 0.5 uniform recombination. Under those settings of P_0, the theory indicates that one-point recombination should perform identically when the second-order hyperplanes have defining length 5.4 and 15, respectively. Since an actual defining length must be an integer, the hyperplanes of defining length 5 and 15 were monitored. It is important to note that this is precisely where the curves of survival (disruption) and construction intersect for one-point and uniform recombination, for second-order hyperplanes (see Fig. 3.9 of Chap. 3 and Fig. 4.6 of Chap. 4).

Figure 10.2 graphs the results. As expected, the results show a perfect match when comparing the evolution of H_2 under 0.5 uniform recombination and one-point recombination when $L_1 = 15$ (the two curves coincide exactly on the graph). The agreement is almost perfect when comparing 0.1

uniform recombination and one-point recombination when $L_1 = 5$, and the small amount of error is due to the fact that the defining length had to be rounded to an integer. As an added comparison, the second-order hyperplanes of defining length 25 were also monitored. The earlier graphs for survival and construction indicate that in this situation one-point recombination should approach equilibrium even faster than 0.5 uniform recombination (because one-point recombination is more disruptive). The graph confirms this observation.

It is important to note that the above analysis holds even for arbitrary cardinality alphabets C, although it was demonstrated for $C = 2$. The system of differential equations would have more equations and terms as C increases, but the computations would still only involve one computation of $P_{\mathrm{d}}(H_k)$ and $P_{\mathrm{c}}(H_k \mid H_m \wedge H_n)$, and those computations would be precisely the same. To see this, consider having $C = 3$, with an alphabet of $\{0, 1, 2\}$. Then #0#0# can be disrupted if recombined with #1#1#, #1#2#, #2#1#, or #2#2#. The probability of disruption is the same as it was above. Similarly it can be shown that the probability of construction is the same as it was above.

What this indicates is that for second-order hyperplanes, the time evolution of two different recombination operators can be compared simply by comparing the graphs for survival (disruption) and construction that were given earlier in Chaps. 3 and 4. Two recombination operators will approach Robbins' equilibrium at the same rate for those situations where their survival (or construction) graphs intersect. Thus the comparative rate at which n-point recombination and P_0 uniform recombination approach equilibrium for second-order hyperplanes is easily determined by inspection.

Unfortunately, this is not as simple to accomplish for higher-order hyperplanes. In these cases the graphs of survival and construction for n-point recombination were averaged uniformly over all the remaining defining lengths $L_2, ..., L_{k-1}$ of H_k (see Chaps. 3 and 4). Although the time evolution of any population undergoing recombination depends on the probabilities of disruption and construction of various lower-order building blocks (as shown above), it will not generally be the case that those building blocks will be uniformly distributed. Thus the simple graphs of survival and construction that were presented in Chaps. 3 and 4 will not necessarily allow one to compare the rate at which n-point and P_0 uniform recombination approach equilibrium for H_k where $k \geq 3$. This is an avenue for future research.

10.3 The Limiting Distribution for Mutation

This section will investigate the limiting distribution of a population of chromosomes undergoing mutation, and will quantify how the mutation rate μ affects the rate at which the equilibrium is approached. Mutation will work on alphabets of cardinality C in the following fashion. An allele is picked for mutation with probability μ. Then that allele is changed to one of the other $C-1$ alleles, uniformly randomly.

Let S be any string of length L: $(a_1, ..., a_L)$. If a population is mutated repeatedly (without selection or recombination) then:

$$\lim_{t\to\infty} p_S{}^{(t)} = \prod_{i=1}^{L} \frac{1}{C} \tag{10.5}$$

where $p_S{}^{(t)}$ is the expected proportion of string S in the population at time t and C is the cardinality of the alphabet.

Equation 10.5 states that a population undergoing only mutation approaches a "uniform" equilibrium distribution in which all possible alleles are uniformly likely at all loci. Thus all strings will become equally likely in the limit. Clearly, since the mutation rate μ does not appear, it does not affect the equilibrium distribution that is reached. Also, the initial population will not affect the equilibrium distribution. However, both the mutation rate and the initial population may affect the *rate* at which the distribution is approached. This will be explored further in the next several subsections.

10.3.1 A Markov Chain Model of Mutation with Cardinality C

To explore the (non-)effect that the mutation rate and the initial population have on the equilibrium distribution, the dynamics of a finite population of strings being mutated will be modeled as follows. Consider a population of P individuals of length L, with cardinality C. Since Geiringer's Theorem (discussed in the last section) focuses on loci, the emphasis once again will be on the L loci. However, since each locus will be perturbed independently and identically by mutation, it is sufficient to consider only one locus. Furthermore, since each of the alleles in the alphabet are treated the same way by mutation, it is sufficient to focus on only one allele (all other alleles will behave identically). Examining only one locus is analogous to considering a pool of P dice, with each die having C faces. Mutation flips a die with probability μ, producing one of the *other* faces uniformly randomly. Since mutation does not favor any particular face (allele), it is sufficient to concentrate just on one particular face.

Let the alphabet be denoted as $\mathcal{A}$ and $\alpha \in \mathcal{A}$ be one of the particular alleles. Let $\overline{\alpha}$ denote all the other alleles. Then define a state to be the number of αs at some locus and a time step to be one generation in which all individuals have been considered for mutation. More formally, let S_t be a random variable that gives the number of αs at some locus at time t. S_t can take on any of the $P+1$ integer values from 0 to P at any time step t. Since this process is memory-less, the transitions from state to state can be modeled with a Markov chain. For an overview of Markov chains, see Winston (1991) and Chap. 12.

The probability of transitioning from state i to state j in one time step will be denoted as $P(S_t = j \mid S_{t-1} = i) \equiv p_{i,j}$. Thus, transitioning from i to j means moving from a state with $S_{t-1} = i$ α's and $P - i$ $\overline{\alpha}$'s to a state with $S_t = j$ α's and $P - j$ $\overline{\alpha}$'s.

Suppose $j \geq i$. This means we are increasing (or not changing) the number of α's. To accomplish the transition requires that $j - i$ more $\overline{\alpha}$'s are mutated to α's than α's are mutated to $\overline{\alpha}$'s. The transition probabilities are:

$$p_{i,j} = \sum_{x=0}^{min\{i,P-j\}} \binom{i}{x} \binom{P-i}{x+j-i} \times \mu^x \left(\frac{\mu}{C-1}\right)^{x+j-i} (1-\mu)^{i-x} \left(1 - \frac{\mu}{C-1}\right)^{P-j-x} \tag{10.6}$$

Let x be the number of α's that are mutated to $\overline{\alpha}$'s. Since there are i α's in the current state, this means that $i - x$ α's are *not* mutated to $\overline{\alpha}$'s. This occurs with probability $\mu^x(1-\mu)^{i-x}$. Also, since x α's are mutated to $\overline{\alpha}$'s then $x + j - i$ $\overline{\alpha}$'s must be mutated to α's. Since there are $P - i$ $\overline{\alpha}$'s in the current state, this means that $P-i-x-j+i = P-x-j$ $\overline{\alpha}$'s are *not* mutated to α's. This occurs with probability $(\mu/(C-1))^{x+j-i}(1-\mu/(C-1))^{P-x-j}$. The combinatorials yield the number of ways to choose x α's out of the i α's, and the number of ways to choose $x + j - i$ $\overline{\alpha}$'s out of the $P - i$ $\overline{\alpha}$'s. Clearly, it isn't possible to mutate more than i α's. Thus $x \leq i$. Also, since it isn't possible to mutate more than $P-i$ $\overline{\alpha}$'s, $x+j-i \leq P-i$, which indicates that $x \leq P - j$. The minimum of i and $P - j$ bounds the summation correctly.

Similarly, if $i \geq j$, we are decreasing (or not changing) the number of α's. Thus one needs to mutate $i-j$ more α's to $\overline{\alpha}$'s than $\overline{\alpha}$'s to α's. The transition probabilities are:

$$p_{i,j} = \sum_{x=0}^{min\{P-i,j\}} \binom{i}{x+i-j} \binom{P-i}{x} \times \mu^{x+i-j} \left(\frac{\mu}{C-1}\right)^{x} (1-\mu)^{j-x} \left(1 - \frac{\mu}{C-1}\right)^{P-i-x} \tag{10.7}$$

The explanation is almost identical to before. Let x be the number of $\overline{\alpha}$'s that are mutated to α's. Since there are $P - i$ $\overline{\alpha}$'s in the current state, this means that $P-i-x$ $\overline{\alpha}$'s are *not* mutated to α's. This occurs with probability $(\mu/(C-1))^x(1-\mu/(C-1))^{P-i-x}$. Also, since x $\overline{\alpha}$'s are mutated to α's then $x+i-j$ α's must be mutated to $\overline{\alpha}$'s. Since there are i α's in the current state, this means that $i - x - i + j = j - x$ α's are *not* mutated to $\overline{\alpha}$'s. This occurs with probability $\mu^{x+i-j}(1-\mu)^{j-x}$. The combinatorials yield the number of ways to choose x $\overline{\alpha}$'s out of the $P - i$ $\overline{\alpha}$'s, and the number of ways to choose $x + i - j$ α's out of the i α's. Clearly, it isn't possible to mutate more than $P - i$ $\overline{\alpha}$'s. Thus $x \leq P - i$. Also, since it isn't possible to mutate more than

i α's, $x + i - j \leq i$, which indicates that $x \leq j$. The minimum of $P - i$ and j bounds the summation correctly.

In general, these equations are not symmetric ($p_{i,j} \neq p_{j,i}$), since there is a distinct tendency to move towards states with a $1/C$ mixture of α's (the limiting distribution). Note also that if $i = j$ both equations give the same transition probabilities, which provides a useful check on the correctness of the above equations:

$$p_{i,i} = \sum_{x=0}^{min\{i,P-i\}} \binom{i}{x} \binom{P-i}{x} \times \mu^x \left(\frac{\mu}{C-1}\right)^x (1-\mu)^{i-x} \left(1 - \frac{\mu}{C-1}\right)^{P-i-x} \tag{10.8}$$

When $0.0 < \mu < 1.0$ all $p_{i,j}$ entries are nonzero and the Markov chain is ergodic. Thus there is a steady-state distribution describing the probability of being in each state after a long period of time. By the definition of steady-state distribution, it cannot depend on the initial state of the system, hence the initial population will have no effect on the long-term behavior of the system. The steady-state distribution reached by this Markov chain model can be thought of as a sequence of P Bernoulli trials with success probability $1/C$. Thus the steady-state distribution can be described by the binomial distribution, giving the probability π_i of being in state i (i.e., the probability that i α's appear at a locus after a long period of time):

$$\lim_{t \to \infty} P(S_t = i) \equiv \pi_i = \binom{P}{i} \left(\frac{1}{C}\right)^i \left(1 - \frac{1}{C}\right)^{P-i} \tag{10.9}$$

Note that the steady-state distribution does not depend on the mutation rate μ or the initial population, although it does depend on the cardinality C. Now Eq. 10.5 states that the equilibrium distribution is one in which all possible alleles are equally likely. Thus the expected number of α's at any locus of the population (at steady state) can be proven to be:

$$\lim_{t \to \infty} E[S_t] = \sum_{i=0}^{P} \binom{P}{i} i \left(\frac{1}{C}\right)^i \left(1 - \frac{1}{C}\right)^{P-i} = \frac{P}{C}$$

To test the theory, the Markov chain for a population of P individuals and mutation rate μ was constructed. The steady-state distribution was then calculated directly from the Markov chain. As a check, the steady-state values were compared to those obtained by using Eq. 10.9 – they agreed in all cases. Figure 10.3 graphs the results, where the cardinality of the alphabet $C = 5$. In the left graph $P = 10$, while in the right graph $P = 20$. Because $C = 5$, the expected number of α's in a column of the population should be two when $P = 10$ and four when $P = 20$. Note how the distributions peak at those

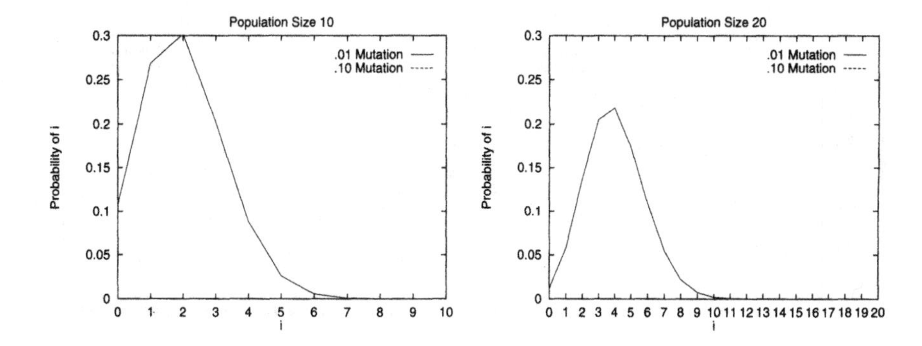

Fig. 10.3. Steady-state distribution for mutation when $C = 5$. $P = 10$ in the left graph and $P = 20$ in the right graph. The mutation rate is irrelevant.

values. Finally, in order to see the effects of the mutation rate, μ was set to 0.01 and 0.10. As expected, changing the mutation rate had no effect on the steady-state distribution (there are actually two curves in each graph – but they are identical).

10.3.2 A Markov Chain Model of Mutation with $C = 2$

When $C = 2$ the equations can be simplified somewhat. In this case the alphabet $\mathcal{A}$ is considered to simply be $\{0, 1\}$. The focus of attention will be on the allele '1' (i.e., $\alpha = 1$). Suppose $j \geq i$. To accomplish the transition requires that $j - i$ more 0s are mutated to 1s than 1s are mutated to 0s.

$$p_{i,j} = \sum_{x=0}^{min\{i,P-j\}} \binom{i}{x} \binom{P-i}{x+j-i} \mu^{2x+j-i}(1-\mu)^{P-(2x+j-i)}$$

Similarly, if $i \geq j$, one needs to mutate $i - j$ more 1s to 0s than 0s to 1s, yielding:

$$p_{i,j} = \sum_{x=0}^{min\{P-i,j\}} \binom{i}{x+i-j} \binom{P-i}{x} \mu^{2x+i-j}(1-\mu)^{P-(2x+i-j)}$$

The steady-state distribution reached by this Markov chain model can be described by giving the probability of being in state i, π_i:

$$\lim_{t\to\infty} P(S_t = i) \equiv \tag{10.10}$$

$$\pi_i = \binom{P}{i} \left(\frac{1}{2}\right)^i \left(1 - \frac{1}{2}\right)^{P-i} = \binom{P}{i} 0.5^P$$

The expected number of 1s at steady state is given by:

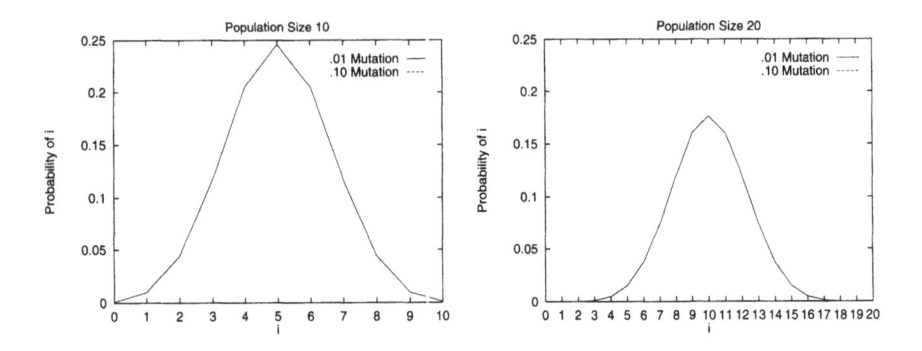

Fig. 10.4. Steady-state distribution for mutation when $C = 2$. $P = 10$ in the left graph and $P = 20$ in the right graph. The mutation rate is irrelevant.

$$\lim_{t\to\infty} E[S_t] = \sum_{i=0}^{P} \binom{P}{i} i\, 0.5^P = \frac{P}{2}$$

To see how the theory is affected by the cardinality of the alphabet, the Markov chain for a population of P individuals and mutation rate μ was constructed when $C = 2$. The steady-state distribution was then calculated directly from the Markov chain – again the results agreed with the theoretically derived distribution given above in Eq. 10.10.

Figure 10.4 graphs the results. In the left graph $P = 10$, while in the right graph $P = 20$. Because $C = 2$, the expected number of 1s in a column of the population should be five when $P = 10$ and ten when $P = 20$. Note how the distributions peak at those values. The distributions are also now symmetric around the mean, because mutation is equally likely to yield an $\alpha = 1$ as an $\overline{\alpha} = 0$. Finally, in order to see the effects of the mutation rate, μ was set to 0.01 and 0.10. As expected, changing the mutation rate had no effect on the steady-state distribution (there are actually two curves in each graph – but they are identical).

10.3.3 The Rate at Which the Limiting Distribution Is Approached

The previous subsections showed that the mutation rate μ and the initial population have no effect on the limiting distribution that is reached by a population undergoing only mutation. However, these factors may very well have an influence on the *rate* at which that limiting distribution is approached. This issue is investigated in this subsection.

In order to model the rate at which the process approaches the limiting distribution, consider an analogy with radioactive decay. In radioactive decay, nuclei disintegrate and thus change state. In the world of binary strings ($C =$ 2) this would be analogous to having a sea of 1s mutate to 0s, or with arbitrary

C this would be analogous to having a sea of α's mutate to $\overline{\alpha}$'s. In radioactive decay, nuclei cannot change state back from $\overline{\alpha}$'s to α's. However, for mutation, states can continually change from α to $\overline{\alpha}$ and vice versa. This can be modeled as follows. Let ${p_\alpha}^{(t)}$ be the expected proportion of α's at time t. Then the expected time evolution of the process can be described by a differential equation:

$$\frac{d{p_\alpha}^{(t)}}{dt} = \tag{10.11}$$
$$-\mu\,{p_\alpha}^{(t)} + \left(\frac{\mu}{C-1}\right)(1-{p_\alpha}^{(t)}) = \left(\frac{\mu}{C-1}\right)(1-C\,{p_\alpha}^{(t)})$$

The term $\mu\,{p_\alpha}^{(t)}$ represents a loss, which occurs if α is mutated. The other term is a gain, which occurs if an $\overline{\alpha}$ is successfully mutated to an α. At steady state the differential equation must be equal to 0, and this is satisfied by ${p_\alpha}^{(t)} = 1/C$, as would be expected.

The general solution to the differential equation was found to be:

$${p_\alpha}^{(t)} = \frac{1}{C} + \left({p_\alpha}^{(0)} - \frac{1}{C}\right) e^{\frac{-C\mu t}{C-1}} \tag{10.12}$$

where $-C\mu/(C-1)$ plays a role analogous to the decay rate in radioactive decay. This solution indicates a number of important points. First, as expected, although μ does not change the limiting distribution, it does affect how fast it is approached. Also, the cardinality C also affects that rate (as well as the limiting distribution itself). Finally, different initial conditions will also affect the rate at which the limiting distribution is approached, but will not affect the limiting distribution itself. For example, if ${p_\alpha}^{(0)} = 1/C$ then ${p_\alpha}^{(t)} = 1/C$ for all t, as would be expected.

The solution can be checked by noting that:

$$\frac{d{p_\alpha}^{(t)}}{dt} = \frac{-C\,\mu\,{p_\alpha}^{(0)}\,e^{\frac{-C\mu t}{C-1}} + \mu\,e^{\frac{-C\mu t}{C-1}}}{C-1}$$

A number of special cases are of interest. The first is when the population is initially seeded only with α's (i.e., ${p_\alpha}^{(0)} = 1$). Then the solution to the differential equation is:

$${p_\alpha}^{(t)} = \frac{(C-1)\,e^{\frac{-C\mu t}{C-1}} + 1}{C}$$

The second case is when the cardinality $C = 2$. Let us assume that binary strings are being used and $\alpha = 1$. Equation 10.11 becomes quite simple:

$$\frac{d{p_1}^{(t)}}{dt} = -\mu\,{p_1}^{(t)} + \mu\,(1-{p_1}^{(t)}) = \mu\,(1-2\,{p_1}^{(t)})$$

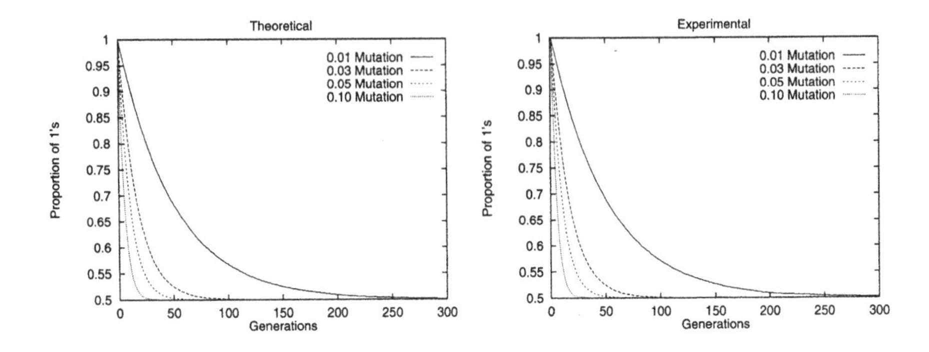

Fig. 10.5. Decay rate for mutation when $C = 2$. The theoretical results are on the left, and experimental results are on the right.

At steady state this must be equal to 0, and ${p_1}^{(t)} = 1/2$. Assume the population is initially seeded only with 1s. Then the solution to the differential equation is:

$$ {p_1}^{(t)} = \frac{e^{-2\mu t} + 1}{2} \tag{10.13} $$

which is similar to the equation derived from physics for radioactive decay.

To test the theory, a population of binary strings was initialized to all 1s and then repeatedly mutated at mutation rate μ. This process was repeated 1000 times to average the results. As time progresses the expected proportion of 1s should approach 50%. Figure 10.5 compares the decay curves derived via Eq. 10.13 (the left graph) with the decay curves derived from this experiment (the right graph). As expected the graphs are identical. The key point is that although μ has no effect on the limiting distribution, increasing μ clearly increases the rate at which that distribution is approached.

10.4 The Limiting Distribution for Mutation and Recombination

The previous sections have considered recombination and mutation in isolation. A population undergoing recombination approaches Robbins' equilibrium, while a population undergoing mutation approaches a uniform equilibrium. What happens when both mutation and recombination act on a population?

The answer is very simple. In general, Robbins' equilibrium is not the same as the uniform equilibrium, hence the population cannot approach both distributions in the long term. In fact, in the long term, the uniform equilibrium prevails and we can state a similar theorem for mutation and recombination. If a population is mutated and recombined repeatedly (without selection):

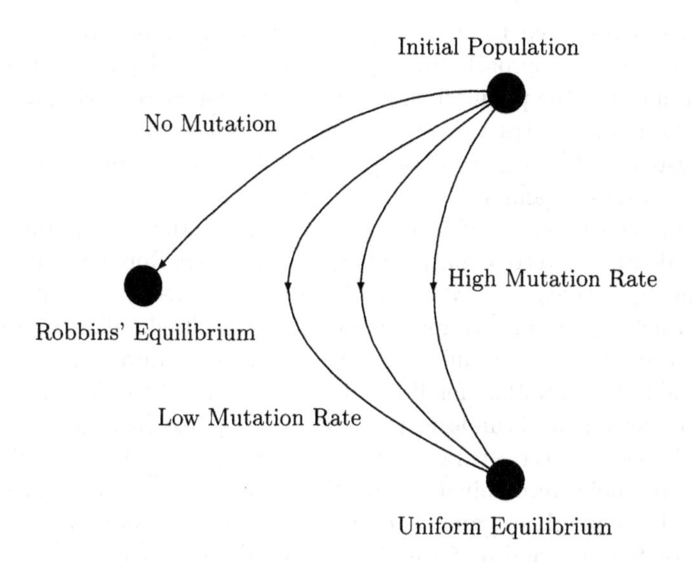

Fig. 10.6. Pictorial representation of the action of mutation and recombination on the initial population

$$\lim_{t \to \infty} p_S{}^{(t)} = \prod_{i=1}^{L} \frac{1}{C} \tag{10.14}$$

This is intuitively obvious. Recombination cannot change the distribution of alleles at any locus – it merely shuffles alleles. Mutation, however, actually changes that distribution. Thus, the picture that arises is that a population that undergoes recombination and mutation attempts to approach a Robbins' equilibrium that is itself approaching the uniform equilibrium. Put another way, Robbins' equilibrium depends on the distribution of alleles in the initial population. This distribution is continually changed by mutation, until the uniform equilibrium distribution is reached. In that particular situation Robbins' equilibrium is the same as the uniform equilibrium distribution. Thus the effect of mutation is to move Robbins' equilibrium to the uniform equilibrium distribution. The speed of that movement depends on the mutation rate μ (greater μ implies faster movement). This is portrayed in Fig. 10.6.

10.5 Summary

This chapter investigated dynamic analyses of recombination and mutation. The dynamic analyses of recombination revolved around the Robbins' equilibrium distribution of strings that will result if a population is repeatedly recombined. Geiringer's Theorem indicates that this equilibrium distribution depends only on the distribution of alleles in the initial population. The form of recombination and the cardinality are irrelevant.

We then attempted to characterize the speed at which various recombination operators approach this equilibrium. By developing a differential equation model of the population, it is possible to show that the probability of disruption and construction of schemata are crucial to the time evolution of the system. Interestingly, these probabilities were obtained from static analyses in Chaps. 3 and 4.

The analysis provides evidence to support the intuitive hypothesis that the more disruptive recombination operators approach Robbins' equilibrium more quickly. Although we were unable to provide precise estimates of the rate at which equilibrium is approached, these results do allow us to make relative statements about different recombination operators. For example, we were able to show that for low-order hyperplanes ($k < 5$), $P_0 = 0.5$ is the setting at which P_0 uniform recombination approaches equilibrium the fastest. Higher or lower settings of P_0 slow the approach. We were also able to compare n-point recombination and P_0 uniform recombination directly on second-order hyperplanes, and we derived a relationship showing when one-point recombination and uniform recombination both approach equilibrium at the same speed. Again, these results hold for arbitrary cardinality. Clearly more work is required to extend these results to higher-order hyperplanes.

We then investigated a dynamic analysis of mutation, and showed that a population undergoing mutation approaches a uniform equilibrium in which every string is equally likely. The mutation rate and the initial population have no effect on that limiting distribution. A differential equation model of this process (which is analogous to radioactive decay in physics) allowed us to compute the speed at which the equilibrium is approached. Both the mutation rate and the initial population affect that speed.

Finally, we investigated the joint behavior of a population undergoing both mutation and recombination. We showed that, in a sense, the behavior of mutation takes priority, in that mutation actually moves Robbins' equilibrium until it is the same as the uniform equilibrium (i.e., all strings equally likely).

All of these dynamic analyses have excluded selection from the process. The remainder of this book will include selection. Chapter 11 will model a population undergoing mutation and selection. Chapter 12 will model a complete evolutionary algorithm with selection, recombination, and mutation.

11. A Dynamic Model of Selection and Mutation

11.1 Introduction

The previous chapters of this book have considered static and dynamic analyses of recombination and mutation, in order to compare the two operators. However, since selection is a crucial component of evolutionary algorithms, it is important to investigate the effects that selection can have on the earlier findings. To this end we now introduce models of evolutionary algorithms that include selection. We will focus primarily on fitness-proportional selection (Holland 1975), due to its mathematical simplicity; however, it would not be difficult to extend the work to other selection mechanisms. This chapter investigates selection and mutation. The next chapter (Chap. 12) investigates selection, mutation, and recombination.

11.2 Selection and Mutation

A population undergoing selection and mutation can be modeled using "equations of motion" that compute the *expected* time evolution of the proportions of the strings (individuals) in the population (equivalently this can be considered to be the evolution of an infinite-size population). In general, since this model will keep track of every possible individual, the model will require a system of C^L simultaneous equations, where L is the string length and C is the cardinality of the alphabet. Let S_i and S_j be arbitrary strings of length L in that alphabet. Let the proportion of a string S_j at time t be denoted as ${p_{S_j}}^{(t)}$. Then the expected time evolution of the system can be computed using the following equation of motion:

$$ {p_{S_j}}^{(t+1)} = \sum_{S_i} {p_{S_i}}^{(t)} \frac{f(S_i)}{\overline{f}^{(t)}} p_{S_i,S_j} \tag{11.1} $$

Equation 11.1 first considers the proportions of all strings S_i at time t. These proportions are modified by fitness-proportional selection, where $f(S_i)$ is the fitness of individual S_i and $\overline{f}^{(t)}$ is the average fitness of the population at time t. Finally p_{S_i,S_j} computes the probability of mutating string S_i to string S_j. The result is the expected proportion of string S_j at time $t+1$.

The total system is described by C^L equations, one for each string S_j. Starting with initial proportions ${p_{S_j}}^{(0)}$, the C^L equations are iterated repeatedly to produce the expected time evolution of the system. Even with binary-string representations ($C = 2$), having binary strings of length $L = 10$ will require over 1000 equations. Clearly this makes it hard to deal with realistic problems. This is a common difficulty that arises when modeling complex systems. Complex systems can often be modeled by using simplifying assumptions. Such models are coarse in the sense that they omit various details of the systems (e.g., our previous models that omitted selection, or the current model that omits recombination). Such models ease analytical burdens but are only approximations to the system. Attempts to increase the fidelity of the approximation by increasing the amount of detail generally worsen the analytical burden. The trick, then, is to find situations under which simplifying assumptions can be made safely. One way to do this is to "aggregate" the system in such a way that multiple elements of the system belong to a particular equivalence class. Then only the equivalence classes need be modeled. If the aggregation is done well, the amount of error introduced into the model may be very small.

It turns out that a surprising number of fitness functions f can be aggregated in a fashion that greatly simplifies the above model by reducing the number of equations and the number of terms in the equations. To see this, let the alphabet be denoted as $\mathcal{A}$ and let $\alpha \in \mathcal{A}$ be one of the alleles. Let $\overline{\alpha}$ denote all the other alleles. It turns out that for some classes of problems (fitness functions), only the number of α's in an individual matters. This form of aggregation introduces *no* error into the model. Thus sets of strings with j α's form an equivalence class, and it suffices to have only $L + 1$ equations, since there can be anywhere from zero to L α's in a string. This is a dramatic reduction from the C^L equations that would be required in the general case. Equation 11.1 appears as before:

$$
{p_{S_j}}^{(t+1)} = \sum_{S_i} {p_{S_i}}^{(t)} \frac{f(S_i)}{\overline{f}^{(t)}} \, p_{S_i,S_j} \tag{11.2}
$$

However, in this case S_i refers to any string with i α's, and S_j refers to any string with j α's. The fitness of any string with i α's is the same, and is denoted as $f(S_i)$. The probability of transitioning from any string with i α's to j α's is given by $p_{S_i,S_j} \equiv p_{i,j}$. Rather than compute these probabilities from scratch, we note that a similar computation in Chap. 10 provided the same probabilities $p_{i,j}$ for strings of length P instead of L. Thus, by simply substituting L for P in Eq. 10.6 we get the following when $j \geq i$:

$$
\begin{aligned}
p_{i,j} = & \sum_{x=0}^{min\{i,L-j\}} \binom{i}{x} \binom{L-i}{x+j-i} \times \\
& \mu^x \left(\frac{\mu}{C-1}\right)^{x+j-i} (1-\mu)^{i-x} \left(1 - \frac{\mu}{C-1}\right)^{L-j-x}
\end{aligned} \tag{11.3}
$$

Similarly, using Eq. 10.7 for $i \geq j$ yields:

$$p_{i,j} = \sum_{x=0}^{min\{L-i,j\}} \binom{i}{x+i-j} \binom{L-i}{x} \times \mu^{x+i-j} \left(\frac{\mu}{C-1}\right)^{x} (1-\mu)^{j-x} \left(1-\frac{\mu}{C-1}\right)^{L-i-x} \quad (11.4)$$

Rather than repeat the explanation of the equations provided in Chap. 10, the reader is urged to consult that chapter for full details.

At first blush, this class of problems would appear to be limited to only unimodal functions (functions with one "peak" in the space). This is not true. For example, consider a familiar two-peak problem where individuals in an EA are binary strings. Traditionally one peak is at 111...111 while the other is at 000...000. However, this problem really monitors only the number of 1s. Suppose there are N 1s. If $N > L/2$ the fitness is N; else the fitness is $L-N$. There are two peaks with maximum fitness L. Many of the deceptive and trap functions investigated by Goldberg (e.g., see Deb and Goldberg 1992) fall in this class. In fact, it is also possible to create problems with an even higher number of peaks. For example, a problem might have a peak where $N = 0$, $N = L/2$, and $N = L$, thus creating a three-peak problem that depends only on one allele. Clearly this technique can be extended to more than three peaks.

Of the unimodal functions mentioned in the literature, two are of interest because they fall within this framework. The first is the class of "Royal Road" functions analyzed in Nimwegen et al. (1997). The fitness function considers each individual to consist of N contiguous blocks of K bits, and the fitness of an individual is simply the number of blocks that consist of K 1s. This is analogous to having an alphabet where $C = 2^K$ in our model, where α is the allele corresponding to the binary string equivalent of K 1s. Interestingly, it appears as if many of the analytical techniques investigated in Nimwegen et al. (1997) could also be applied to the more general class of problems defined here.

The second function is from a widely studied problem in the biological community (Muller 1964). This problem is such that the fitness of an individual is $(1-S)^M$, where M is the distance of the individual from some optimum and S is a selection pressure. Thus the optimal individual has fitness 1.0 ($M = 0$), while nonoptimal individuals have positive fitness less than 1.0.

Since the latter problem was especially designed to investigate the effects of mutation and selection on evolution, we use it as a test function for our mathematical framework. The mathematical model (consisting of Eqs. 11.2–11.4) was compared to the behavior of a standard EA (with recombination turned off) on the $(1-S)^M$ function. Binary strings of length 64 were initialized to all 1s (which is the optimum string), and then the system was allowed

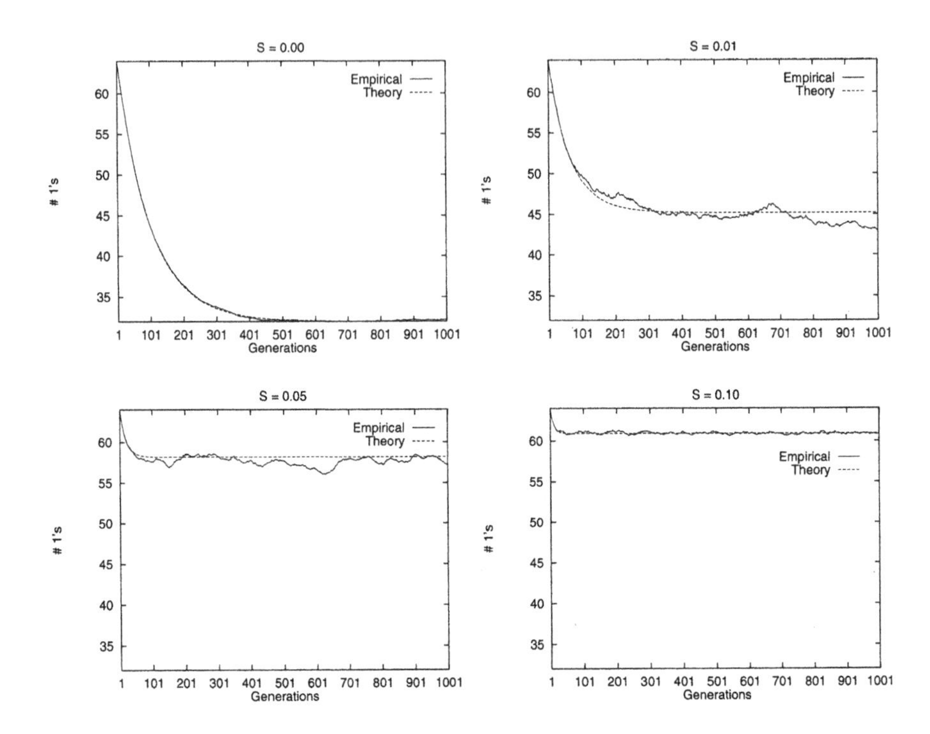

Fig. 11.1. Selection and mutation modeled when $\mu = 0.005$. S takes on values 0.00, 0.01, 0.05, and 0.10.

to evolve. The population size of the EA was 1000. The average number of 1s in the strings was monitored for 1000 generations. Since binary strings are used the distance metric is Hamming distance. Figure 11.1 graphs the results when the mutation rate μ is 0.005 as the selection pressure S ranges from 0.00 to 0.10. Figure 11.2 graphs the results when the mutation rate μ is 0.01 as the selection pressure S ranges from 0.00 to 0.10. Both the theoretical and empirical curves are plotted. Although there is some noise due to stochastic fluctuations in the EA (the EA was only run once per problem), the agreement between theory and experiment is quite good. Note that higher mutation rates drive the system to the same equilibrium distribution, but more quickly, when $S = 0.00$. In fact, when $S = 0.00$, there really is no selection occurring, and the results from Chap. 10 (concerning the limiting distribution of a population undergoing mutation) hold. For example, the expected proportion of 1s, ${p_1}^{(t)}$, in the individuals is governed by Eq. 10.13:

$$ {p_1}^{(t)} = \frac{e^{-2\mu t} + 1}{2} $$

For $S > 0$, selection comes into play. Two observations are immediately obvious from the graphs. The first is that higher mutation rates actually

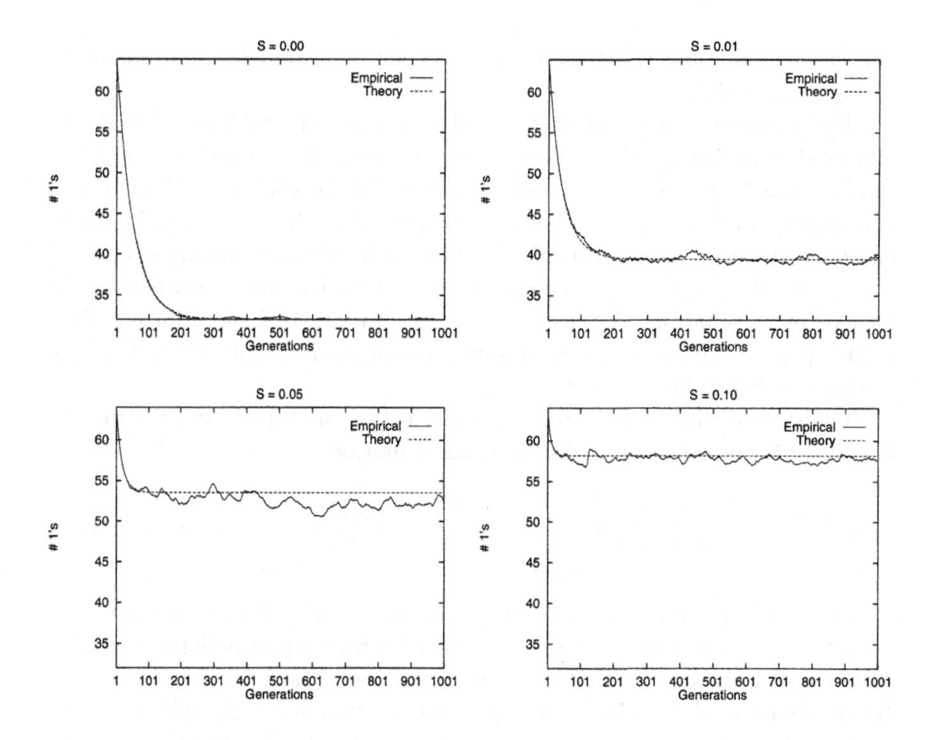

Fig. 11.2. Selection and mutation modeled when $\mu = 0.01$. S takes on values 0.00, 0.01, 0.05, and 0.10.

change the equilibrium distribution, producing strings with less 1s. The second is that higher selection pressure also changes the equilibrium distribution, producing strings with more 1s. Both of these results are intuitively reasonable.

11.3 Summary

This chapter described a model of mutation and fitness-proportional selection in which equations of motion give the expected proportion of the strings in the population over time. In the worst case this involves C^L simultaneous equations. However, we have defined a class of fitness functions that allows for an error-free aggregation of the model that results in far fewer equations. If a fitness function depends only on one particular allele in the alphabet, only $L+1$ equations are required. We pointed out that this class of functions includes common one-peak functions from the literature (the Royal Road function and the $(1-S)^M$ function from the biological community). It also includes many of the two-peak trap and deceptive functions that have been analyzed in the EA community (e.g., see Deb and Goldberg 1992). Finally, this class of functions can also include problems of even higher multimodality (number of peaks).

The main advantage of this model and class of problems is that the expected behavior of reasonably large problems (i.e., problems in high-dimensionality spaces) can be theoretically evaluated with a small number of simultaneous equations. Another advantage is that the model can be easily extended to cover other forms of selection (such as linear-ranking selection), while still allowing the same aggregation to occur. The main drawback is the lack of recombination. However, since some varieties of evolutionary algorithms (e.g., evolutionary programming) do not use recombination, this style of analysis can in fact be quite useful.

The system with fitness-proportional selection, mutation, and recombination can also be modeled with equations of motion:

$$p_{S_j}{}^{(t+1)} = \sum_{S_h} \sum_{S_l} \sum_{S_i} p_{S_h}{}^{(t)} \, p_{S_l}{}^{(t)} \, \frac{f(S_h)}{\overline{f}^{(t)}} \, \frac{f(S_l)}{\overline{f}^{(t)}} \, p_{(S_h \times S_l), S_i} \, p_{S_i, S_j} \quad (11.5)$$

Equation 11.5 first considers the proportions of all pairs of strings S_h and S_l at time t. These proportions are modified by fitness-proportional selection, yielding the probability that S_h and S_l will be chosen for recombination. Recombination of S_h and S_l will produce an individual S_i, which can be mutated into string S_j. The term $p_{(S_h \times S_l), S_i}$ is the probability that the two strings S_h and S_l will be recombined to create S_i, while p_{S_i, S_j} computes the probability of mutating S_i into S_j. The third summation considers all possible strings S_i. The result is the expected proportion of string S_j at time $t + 1$. Once again, the total system is described by C^L equations, one for each string S_j. Starting with initial proportions $p_{S_j}{}^{(0)}$, the C^L equations are iterated repeatedly to produce the expected time evolution of the system.

One interesting observation is that the probability of recombination, $p_{(S_h \times S_l), S_i}$, can be derived using the earlier static schema analyses in Chaps. 3 and 4. Suppose that either S_h or S_l has the correct allele for S_i at every locus (if this is not true, then the probability of creating S_i is 0). Suppose that S_h and S_l differ at $\Delta(S_h, S_l) \equiv k$ alleles. For example, the string AAAA can (potentially) be constructed from the two strings AABB and ABAA (where $k = 3$). Of the k differing alleles, m are at string S_h and $n = k - m$ are at string S_l. Thus what is happening is that two nonoverlapping, lower-order building blocks H_m (in S_h) and H_n (in S_l) are being constructed to form H_k (and thus the string S_i). This occurs with probability:

$$p_{(S_h \times S_l), S_i} = P_c(H_k \mid H_m \wedge H_n) \qquad \text{where } 0 \le \Delta(S_h, S_l) \equiv k \le L$$

which is equivalent to the probability of construction that was introduced in the static schema analysis performed in Chap. 4.[1]

[1] If $k = 0$ or $k = 1$ then either $S_h = S_i$ or $S_l = S_i$ and this is survival, not construction (see Chap. 3).

The main drawback of this model is that the addition of recombination dramatically worsens the complexity of each of the C^L equations. A similar difficulty can be seen in Whitley's "executable model" (Whitley 1992), which is similar in spirit, although it lacks mutation. Also, the addition of recombination makes aggregation much more difficult. For example, since recombination is affected by the location of alleles on an individual, monitoring the number of some allele (as was done above to aggregate the system with selection and mutation) will be problematic.[2]

The second drawback is the assumption of an infinite population, which does not capture finite-population effects. In order to overcome this drawback the next chapter (Chap. 12) investigates a Markov model of a complete EA, with a finite population that undergoes fitness-proportional selection, mutation, and recombination. Unfortunately, the assumption of a finite population complicates the model even further, making aggregation even more necessary. To deal with this issue, Chap. 13 gives a novel algorithm for automatically aggregating any Markov model.

[2] However, Annie Wu (personal communication) has shown that it may be possible to estimate $p_{(S_h \times S_l),S_i}$ for the "one allele" equivalence class, where the two parents have h and l α's and the offspring has i α's.

12. A Dynamic Model of Selection, Recombination, and Mutation

12.1 Introduction

The static (Chaps. 3–9) and dynamic (Chap. 10) theories that have been used to analyze mutation and recombination have omitted selection. These theories were relatively easy to analyze, and they provided useful insights into the various components of an EA. However, because they do not include selection, they are generally insufficient as predictive theories. Chapter 11 included selection, and modeled the behavior of an *infinite*-population EA with selection and mutation (with and without recombination). It turns out that the model without recombination can be efficiently analyzed on a restricted (but interesting) class of functions. The addition of recombination makes the model considerably more complex.

This turns out to be a general observation – the more complete the model of an EA, the more predictive it becomes, with the commensurate increased cost in analytical complexity. This chapter will conclude the trend towards completeness, by examining a complete model of an EA with a *finite* population, selection, recombination, and mutation.

12.2 EA Performance

What sorts of questions should a predictive theory of EAs be able to answer? The standard measures of performance for optimization algorithms involve convergence properties (i.e., the ability to find an optimum) as well as convergence rates (i.e., how quickly an optimum is found). Since EAs are parallel, population-based, stochastic search procedures, there are a number of possible definitions of convergence. The simplest notion is that ultimately an EA population converges to a uniform population consisting of P copies of a single individual which may or may not correspond to a global optimum.

Since most EAs are run with nonzero mutation rates, this simple form of convergence seldom occurs, unless one "anneals" the mutation rate over time (Fogel 1995). Without an annealing mechanism, EAs settle into a dynamic equilibrium in which the exploratory pressures of mutation and recombination are balanced by the exploitative pressure of selection (for example, this can be seen in Chap. 11 for an EA with selection and mutation). Moreover, if mutation is active, every point in the space has some nonzero probability of being visited (infinitely often). Hence, it is trivial to show that a global optimum will be visited infinitely often when an EA is left to run in this state of dynamic equilibrium.

As a consequence, most EA practitioners measure performance in terms of the average (or best) points in the current population, or in terms of monotonically nondecreasing "best-so-far" curves which plot, as a function of the number of samples (or generations), the best point found so far in the search process regardless of whether or not that point is currently represented in the population.

Some natural questions related to such performance measures immediately arise. How likely is it that, if I look at the contents of the nth generation, it will contain a copy of the optimum? What is the expected waiting time until a global optimum is encountered for the first time? How long does it take before a point is encountered that is within some error tolerance of the optimum? How much variance is there in such measures from run to run? How much are these measures affected by changes in population size, mutation rates, etc.?

The goal of this chapter is to provide a framework for answering the previous questions concerning the behavior of EAs. Such a predictive theory must simultaneously take into account the characteristics of the particular EA being used (generational, elitist, etc.), the internal search space representation (binary, gray code, etc.), the operators used (form and rate of recombination, etc.), the nonlinear dynamics of the search process, and the characteristics of the function to be optimized.

12.3 Overview of Markov Chains

If statistics such as the mean and variance of waiting times are to be used as measures of performance, random process theory would seem to provide an appropriate set of tools for describing the behavior of stochastic EAs. Historically, it has been quite natural to model simple EAs as Markov chain processes in which the "state" of an EA is given by the contents of the current population (De Jong 1975; Goldberg and Segrest 1987). One can then imagine a state space of all possible populations and study the characteristics of the population trajectories (the Markov chains) an EA produces from randomly generated initial populations. This is the approach taken in this chapter.

12.3.1 Basic Definitions

A discrete-time Markov chain is a dynamical system composed of N discrete states. At each time step, the Markov chain can change states. Let S_t be the random variable for the Markov chain, which can take on any of the N states at time t. The system is described by an $N \times N$ matrix Q, which gives the probability of transitioning from one state i at time $t-1$ to state j at time t:

$$Q(i,j) \equiv p_{i,j} \equiv P(S_t = j \mid S_{t-1} = i)$$

The $p_{i,j}$ values define the "one-step" probability transition matrix Q, since Q describes the probability of transitioning from state to state in one time step. The transient behavior of the system is obtained from the "n-step" probability transition values, which are obtained from the nth power of Q:

$$Q^n(i,j) \equiv {p_{i,j}}^{(n)} \equiv P(S_t = j \mid S_{t-n} = i)$$

It is also possible to compute conditional probabilities over a set of states. Define a predicate $Pred_J$ and the set J of states that make $Pred_J$ true. Then the probability that the system will be in one of the states of J at time t, given it is in state i at time $t-n$ is:

$${p_{i,J}}^{(n)} \equiv P(S_t \in J \mid S_{t-n} = i) = \sum_{j \in J} {p_{i,j}}^{(n)}$$

Analyses of Markov chain behavior can be considered to be "instantaneous" or "cumulative." Instantaneous behavior refers to events that occur *at* a particular time. An example would be whether the system is at a particular state at that time. Cumulative behavior refers to events that have occurred *by* a certain time. This chapter will define both sets of behavior, but will focus primarily on instantaneous behavior.

12.3.2 Instantaneous Transient Behavior at Time n

The probability of being in some state j at time n is given by simply considering the probability of each possible n-step transition, appropriately weighted by the a priori probabilities:

$${p_j}^{(n)} \equiv P(S_n = j) = \sum_i {p_i}^{(0)} \, {p_{i,j}}^{(n)}$$

where the a priori probability of a system being in state i at time 0 is denoted as ${p_i}^{(0)}$.

The probability that the system is in one of the states of J at time n is:

$${p_J}^{(n)} = \sum_{j \in J} {p_j}^{(n)} \qquad (12.1)$$

Finally, it is possible to compute the probability that the system will transition from one set of states to another set of states. Let $Pred_I$ be another predicate over the states, and denote I to be the set of states that make $Pred_I$

true. Then the probability that the system will be in one of the states of J at time t, given that it is in one of the states of I at time $t-n$, is:

$$p_{I,J}{}^{(n)} \equiv \qquad (12.2)$$
$$P(S_t \in J \mid S_{t-n} \in I) = \frac{\sum_{i \in I} p_i{}^{(t-n)} \; p_{i,J}{}^{(n)}}{p_I{}^{(t-n)}} = \frac{\sum_{i \in I} p_i{}^{(t-n)} \; p_{i,J}{}^{(n)}}{\sum_{i \in I} p_i{}^{(t-n)}}$$

which involves a renormalization over the states indexed by I. Since Eq. 12.2 describes how a system transitions from a group of states to another group of states, this raises the intriguing notion that a system with a large number of states might be simplified (or aggregated) into a system with a smaller number of groups of states. Unfortunately, this is hard to do in general since Eq. 12.2 is a time-dependent equation (thus making the aggregated Markov chain nonstationary), so this discussion will be deferred until Chap. 13.

The nice feature of this formalization is that *any* predicate over the states (populations) can be used. Thus, if the system is an EA, and the focus is on optimality, it is natural to define the set of states that contain at least one copy of an optimum, and compute the probability that the EA will actually be in one of these states at generation n. Predicates that select states based on average fitness, fitness variance, population homogeneity, and so on, can also be of interest.

It is possible to generalize further to arbitrary functions f over the states and compute, for example, the expected value of that function, at time n:

$$E[f]^{(n)} = \sum_j p_j{}^{(n)} \; f(j) \qquad (12.3)$$

12.3.3 Cumulative Transient Behavior

Another common computation involving Markov chains is referred to as the "mean first passage times" for going from state i to state j (for a nice discussion of this, see Winston 1991). This refers to the length of time that it takes (on the average) to reach state j for the first time, given that the process is currently in state i. Answering such questions involves solving the set of simultaneous equations:

$$m_{i,j} \equiv p_{i,j} + \sum_{k \neq j} p_{i,k} \; (1 + m_{k,j}) \qquad (12.4)$$

where $m_{i,j}$ denotes the mean first passage time from state i to state j. To understand the equation, consider transitioning from state i to j in one move. This occurs with probability $p_{i,j}$ and requires only one step. However, suppose the system transitions from state i to state k, where k is not equal to j. This occurs with probability $p_{i,k}$ and requires one step. However, there now remain $m_{k,j}$ steps to state j.

As before, if there is interest in a set J of states, it is possible to compute the mean first passage time for the system to first enter that set of states, given that it is currently outside that set:

$$m_{i,J} \equiv \sum_{j \in J} p_{i,j} + \sum_{k \notin J} p_{i,k} \, (1 + m_{k,J})$$

where $m_{i,J}$ denotes the mean first passage time from state i to any of the states in set J, and i is not in J. This is very similar to Eq. 12.4, with the exception that the probability of entering state J in one step is simply the sum of the probabilities of entering each state within J.

Once this system of simultaneous equations is solved, it is possible to calculate the "expected waiting time" to reach a state in J, given a random initial state, via:

$$EWT_J = \sum_{i \in J} {p_i}^{(0)} \, 0 + \sum_{i \notin J} {p_i}^{(0)} \, m_{i,J}$$

There are two parts to this equation. The first part reflects the possibility that a random initial population is in state J, and hence has a zero waiting time. The second part reflects the mean passage time from initial populations not in J, to a state in J. Clearly this simplifies to:

$$EWT_J = \sum_{i \notin J} {p_i}^{(0)} \, m_{i,J}$$

This holds for any set of states J, and thus it can be used to provide expected waiting times for a variety of events.

12.4 The Nix and Vose Markov Chain Model for EAs

Most of the analytic results from the prior Markov chain approaches for EAs are derived using infinite-population models and involve characterizing steady-state behavior (Davis and Principe 1991; Vose 1992; Suzuki 1993; Rudolph 1993). It is considerably more difficult to get analytic results concerning transient behavior, such as the means and variances of waiting times, for Markov models of finite-population EAs. However, increases in computer technology now permit the visualization and computational exploration of such models as the first steps in developing such a theory. Among the many papers on Markov models of EAs, the Nix and Vose model (1992) is particularly well suited to serve as the basis for the framework provided in this chapter.

The Nix and Vose Markov model is intended to represent a simple, generational EA consisting of a finite population, a standard binary representation,

Table 12.1. The number of states N as a function of L and P

	String Length L				
P	1	2	3	4	5
1	2	4	8	16	32
2	3	10	36	136	528
3	4	20	120	816	5,984
4	5	35	330	3,876	52,360
5	6	56	792	15,504	376,992
6	7	84	1,716	54,264	2,324,784
7	8	120	3,432	170,544	12,620,256
8	9	165	6,435	490,314	61,523,748
9	10	220	11,440	1,307,504	273,438,880
10	11	286	19,448	3,268,760	1,121,099,408

standard mutation and recombination operators, and fitness-proportional selection. Fitness scaling, elitism, and other optimization-oriented features are not modeled.

If L is the length of the binary strings, then $r = 2^L$ is the total number of possible strings. If P is the population size, then the number of possible populations, N, corresponding to the number of possible states, is given in Nix and Vose (1992):

$$N = \binom{P+r-1}{r-1} \tag{12.5}$$

Unfortunately, the size of the $N \times N$ matrix Q for typical EA applications is computationally unmanageable since the number of states N grows rapidly with population size P and string length L (see Eq. 12.5 and Table 12.1). However, as will be found, initial results from models involving small values of P and L can hold as the model scales to more realistic sizes.

The possible populations are described by the matrix Z, which is an $N \times r$ matrix. The ith row $\phi_i = \ < z_{i,0}, \ldots, z_{i,r-1} >$ of Z is the incidence vector for the ith population. In other words, $z_{i,y}$ is the number of occurrences of string y in the ith population, where y is the integer representation of the binary string. For example, suppose $L = 2$ and $P = 2$. Then $r = 4$, $N = 10$, and the Z matrix is shown in Table 12.2.

Nix and Vose then define two mathematical operators, $\mathcal{F}$ and $\mathcal{M}$, where $\mathcal{F}$ is determined from the fitness function, and $\mathcal{M}$ depends on the mutation rate μ, recombination rate χ (this is the percentage of individuals that are recombined every generation), and form of recombination and mutation used. In their paper they assume a standard bit flipping mutation operator and a one-point recombination which produces a single offspring, although $\mathcal{M}$ can be generalized to other operators. With $\mathcal{F}$ and $\mathcal{M}$ defined, they are now able to calculate exact state transition probabilities $p_{i,j}$ via:

Table 12.2. The Z matrix when $L = 2$ and $P = 2$

	Binary String			
State	00	01	10	11
1	0	0	0	2
2	0	0	1	1
3	0	0	2	0
4	0	1	0	1
5	0	1	1	0
6	0	2	0	0
7	1	0	0	1
8	1	0	1	0
9	1	1	0	0
10	2	0	0	0

$$Q(i,j) = p_{i,j} = P! \prod_{y=0}^{r-1} \frac{\mathcal{M}\left[\frac{\mathcal{F}\ \phi_i}{|\mathcal{F}\ \phi_i|}\right]_y^{z_{j,y}}}{z_{j,y}!} \tag{12.6}$$

That is, given $\mathcal{F}$ and $\mathcal{M}$, $p_{i,j}$ specifies how likely it is that a simple EA in state i (the current population) will be in state j in the next generation. One can see from the equation that fitness-proportional selection is assumed.

If the mutation rate μ is nonzero, all states have some nonzero probability of being reached. Hence all the entries of Q are nonzero making the Markov chain ergodic. It is a theorem that any ergodic Markov chain has a limiting distribution called the "steady-state distribution." This implies that, in the limit of many generations (time steps), the probability of being in any state does not depend on the starting state of an EA.

Naturally, most of the interesting behavior for an EA is the transient behavior, which occurs before the steady-state distribution is reached. One way to investigate transient behavior is through the visualization of the n-step probability transition matrices Q^n. This is done in De Jong et al. (1994) as well as in Spears and De Jong (1996), but will not be addressed in this chapter. Instead, this chapter will concentrate on the computational techniques for exploring the transient behavior of the EA. The focus will be on the instantaneous transient behavior (for example, as computed by Eq. 12.1 or Eq. 12.3). Experiments concerning the cumulative transient behavior (i.e., expected waiting time analyses) can be found in De Jong et al. (1994) and in Rees and Koehler (1999).

12.5 Instantaneous Transient Behavior of EAs

While theorems can be proven regarding the long-term steady-state behavior of an EA, they don't directly answer the questions raised earlier, such as how likely is it that the optimum will be present in the nth generation.

Answering such questions requires the computation of the transient behavior of the Markov chain (i.e., the time *before* steady-state behavior is reached). For example, it would be useful to answer the following questions that focus on instantaneous behavior at generation n:

1) What is the probability that an EA population will contain a copy of the optimum at generation n?
2) What is the probability that an EA population will have average fitness greater than some value at generation n?
3) What is the probability that an EA population will have homogeneity less than some value at generation n?
4) What is the expected best individual at generation n?

To answer such questions, it is necessary to combine Q^n with a set of initial conditions concerning an EA at generation 0. For this chapter it is assumed that EA populations are randomly initialized. Thus, the a priori probability of being in state i at time 0, denoted as ${p_i}^{(0)}$, is:

$$ {p_i}^{(0)} = \frac{P!}{z_{i,0}! \ldots z_{i,r-1}!} \left[\frac{1}{r} \right]^P \tag{12.7} $$

Since there are $r = 2^L$ possible strings, each string has a probability of r^{-1} of occurring. The power P takes into account that there are P strings in the population, and the multinomial distribution takes into account the different ways the strings can be inserted into the population to create a unique state. Now it is possible to answer the four questions posed at the beginning of this section:

1) To compute the probability that an EA will have in the population at time n at least one copy of the optimum, use Eq. 12.1 with J as the set of all populations containing at least one copy of the optimum.
2) To compute the probability that an EA will have at time n a population with an average fitness greater than X, use Eq. 12.1 with J as the set of all populations having average fitness greater than X.
3) To compute the probability that an EA will have at time n a population with homogeneity less than X, use Eq. 12.1 with J as the set of all populations having homogeneity less than X.
4) To compute the expected best fitness value in the population at time n, use Eq. 12.3 with f defined to return the maximum fitness in a given population.

We are now in a position to analyze the behavior of a finite-population EA on particular classes of fitness functions. Since the emphasis in the book is on the role of recombination and mutation in EAs, we will focus our attention on a class of problems designed to be difficult for recombination, namely "multimodal" functions. The motivation and design of these functions stems from our earlier schema analyses provided in the book.

12.5.1 Instantaneous Transient Behavior and Multimodality

Consider the following four functions, defined in Table 12.3, where $L = 2$ and the optimum string is always at "11." The four functions differ only in the fitness value of the string "00." In the first function, the further a string is from the optimum string "11," the lower the fitness. Thus this can be considered to be a "one-peak" function. On the other hand, the fourth function can be considered to be a "two-peak" function, with a local suboptimum at "00." The second and third functions fall in between, as the fitness of the local optimum at "00" is changed.

Table 12.3. Four functions where $L = 2$

	Fitness Function			
Funct.	f(00)	f(01)	f(10)	f(11)
1	0.01	0.1	0.1	4.0
2	0.1	0.1	0.1	4.0
3	1.0	0.1	0.1	4.0
4	2.0	0.1	0.1	4.0

The motivation for examining this set of functions stems from the schema theory presented earlier in this book. As pointed out in Chap. 8, recombination will be most useful when high-fitness building blocks of relatively high order (H_m and H_n) can be combined into higher-order building blocks (H_k) that are also of high fitness. Recombination will be least useful when the higher-order building blocks that are constructed have poor fitness. Although schema theory is not strictly predictive, the expectation from this theory is that recombination should help on the first function, since the medium-fitness schemata #01# and #10# can be recombined to create the optimum schema #11#. However, for the fourth function, the two schemata #01# and #10# now have (relatively) low fitness. Furthermore the medium-fitness schema #00# cannot be recombined with any other schema to create the optimum schema #11#. This would appear to be much more difficult for recombination. Again, the second and third functions represent intermediate functions. The expectation is that recombination should perform worse and worse as one proceeds from the first to fourth function.

Consider plotting the probability that the EA will have a copy of the optimum in its population at generation n as n increases, on all four functions (using Eq. 12.1). Figure 12.1 illustrates the results (using the Nix and Vose Markov chain model), in which one-point recombination is turned off ($\chi = 0.0$) and on ($\chi = 1.0$) while holding the mutation rate fixed at $\mu = 0.1$. The population size P was five. Recall that there is no elitism.

In comparing the performance of recombination and mutation on the first and fourth functions (in Fig. 12.1) one can see some confirmation of our expectations. On the first function (one peak) the complete EA with recom-

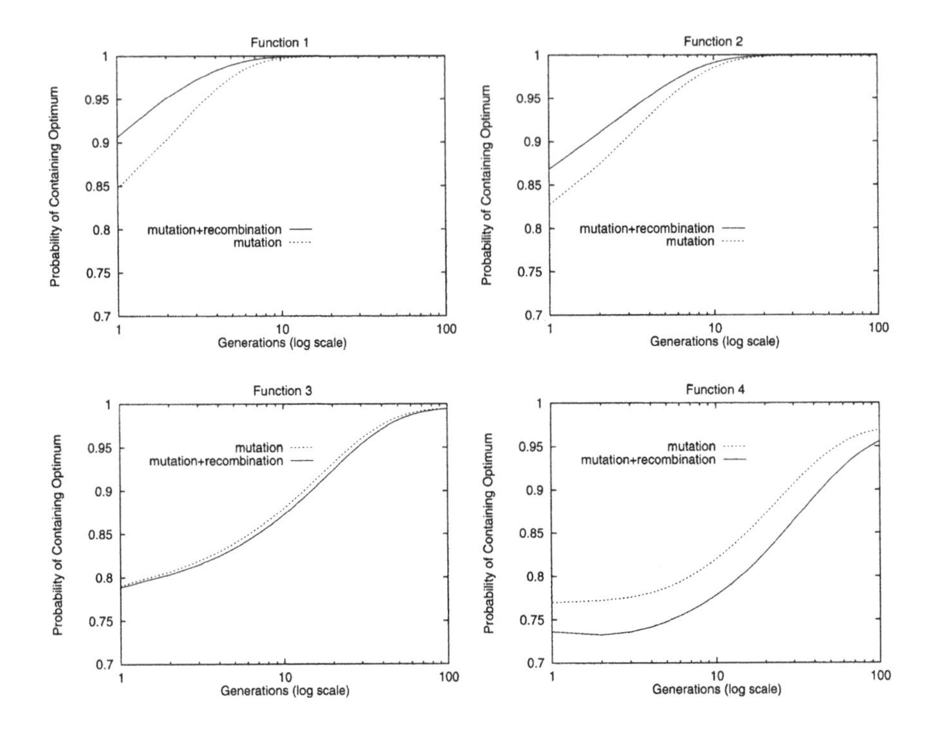

Fig. 12.1. EA behavior on the four functions, where $P = 5$, $L = 2$, and $\mu = 0.1$. The behavior is the probability of containing the optimum at each generation.

bination is better than the EA without recombination, while on the fourth function the EA without recombination is better. These results are consistent with our expectations from the schema theory, and thus support the intuition that recombination can exploit useful building blocks when they are present, but can actually degrade performance when they are not present (or when the higher-order building blocks that are constructed via recombination have poor fitness). One tentative conclusion then is that the number of peaks in the fitness landscape can indeed have a significant influence on the relative performance of recombination in an EA. When there are multiple peaks, recombination is likely to recombine individuals on different peaks, creating low-fitness offspring.

Upon comparing performance on all four functions, one can see that the performance of recombination smoothly degrades as the fitness of the string "00" increases. Thus, another tentative conclusion is that both the number of peaks and their fitness are important factors influencing the performance of recombination. Recombination appears to perform worst when the peaks have a similar height (maximum fitness). However, as the fitness of suboptimal peaks is reduced, the probability of recombining individuals on the same (high) peak increases, increasing the effectiveness of recombination.

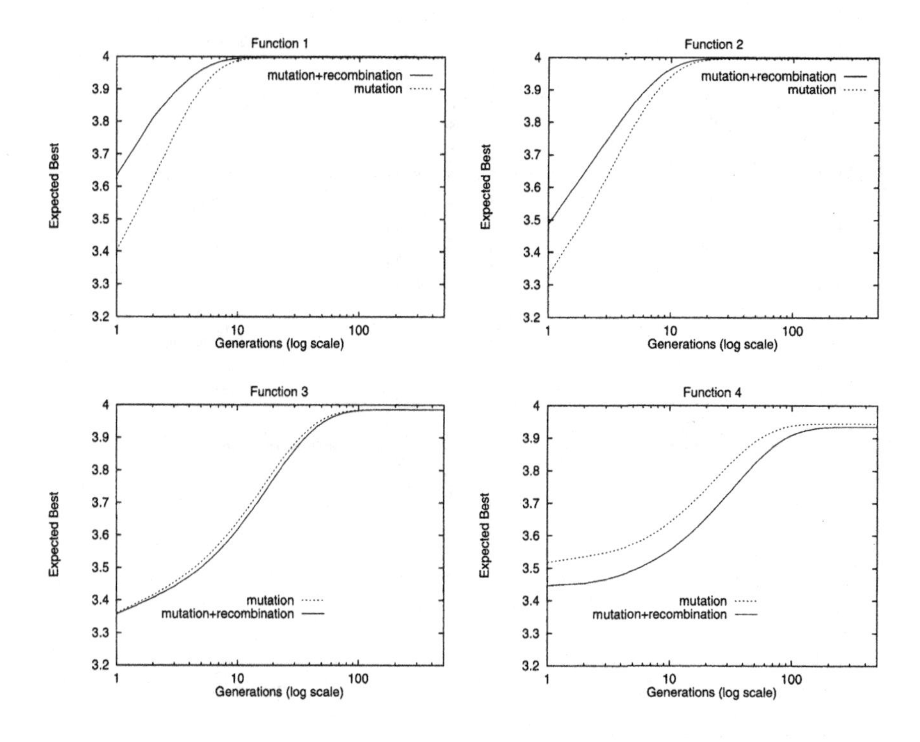

Fig. 12.2. EA behavior on the four functions, where $P = 5$, $L = 2$, and $\mu = 0.1$. The behavior is the expected best individual at each generation.

Table 12.4. Four functions where $L = 3$

Funct.	Fitness Function							
	f(000)	f(001)	f(010)	f(011)	f(100)	f(101)	f(110)	f(111)
1	0.01	0.1	0.01	0.1	0.1	3.0	0.1	4.0
2	0.1	0.1	0.1	0.1	0.1	3.0	0.1	4.0
3	1.0	0.1	1.0	0.1	0.1	3.0	0.1	4.0
4	2.0	0.1	2.0	0.1	0.1	3.0	0.1	4.0

As opposed to monitoring the probability of seeing the optimum at generation n, a more traditional performance measure is to monitor the expected best fitness seen at each generation n. Consider plotting the expected best fitness at generation n as n increases, on all four functions (using Eq. 12.3 where the function f returns the fitness of the best individual in every population). Figure 12.2 illustrates the results, in which recombination is turned off ($\chi = 0.0$) and on ($\chi = 1.0$) while holding the mutation rate fixed at $\mu = 0.1$. As can be seen, the results mirror those in Fig. 12.1.

To see whether this behavior scales to larger problems, consider the following four functions, defined in Table 12.4, where $L = 3$ and the optimum string is always at "111." The four functions differ only in the fitness value

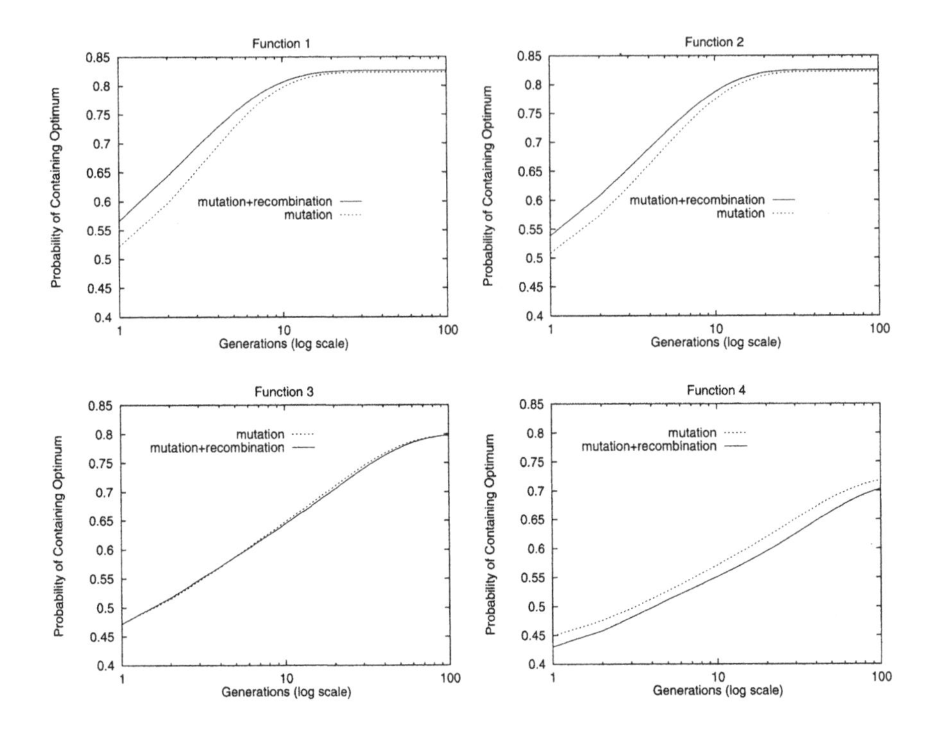

Fig. 12.3. EA behavior on the four functions, where $P = 4$, $L = 3$, and $\mu = 0.1$. The behavior is the probability of containing the optimum at each generation.

of the strings "000" and "010." Again, the first function can be considered to be a one-peak function, while the fourth function can be considered to be a two-peak function, with a local suboptimum at "000" and "010." The second and third functions fall in between, as the fitness of the local optimum is changed.

Figures 12.3 and 12.4 illustrate the behavior of the EAs, in which recombination is turned off ($\chi = 0.0$) and on ($\chi = 1.0$) while holding the mutation rate fixed at $\mu = 0.1$. The population size P was four. Once again the behavior is as expected. Recombination performs worst on the two-peak problem and improves as the height of the local optimum is decreased, until one reaches the one-peak problem.

Thus, it appears as if the number of peaks in a space (the multimodality), as well as the relative heights of those peaks, will provide a useful mechanism for investigating the relative usefulness of recombination and mutation. In order to investigate this observation further, Chap. 14 will introduce the notion of "test-problem generators" that can create random problems in which the number of peaks (and their fitness) can be controlled. The results obtained with a real EA on these problems will validate the observations made from the Markov chain approach performed in this chapter.

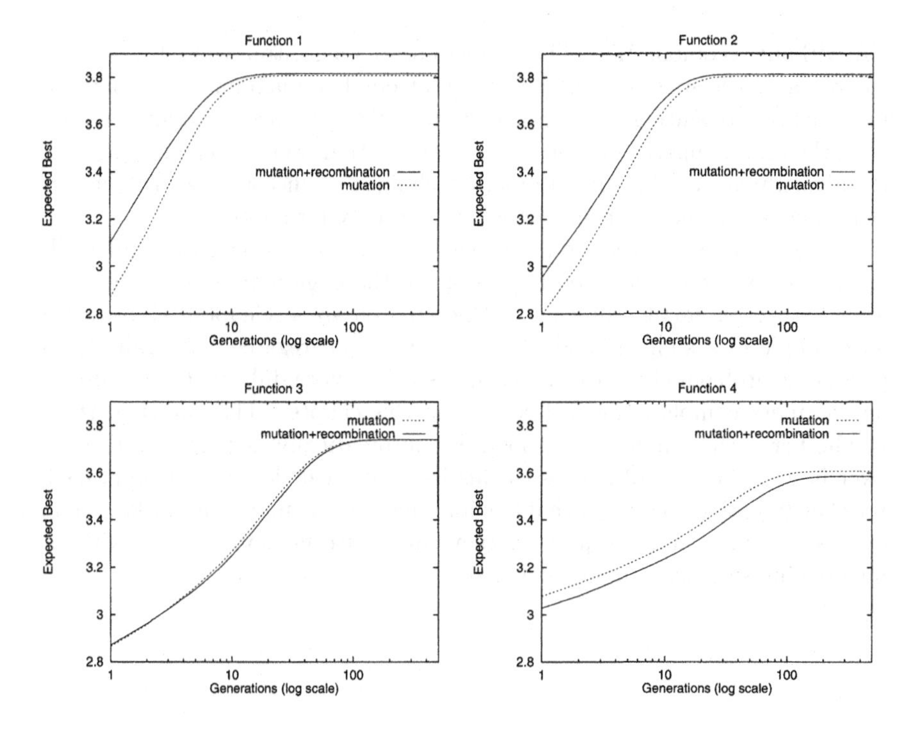

Fig. 12.4. EA behavior on the four functions, where $P = 4$, $L = 3$, and $\mu = 0.1$. The behavior is the expected best individual at each generation.

12.6 Summary and Discussion

This chapter describes some initial explorations of a transient Markov chain analysis as the basis for a stronger EA theory. Although closed-form analysis is difficult in general, useful insights can be obtained by means of the computational exploration of the transient behavior of the models. The initial progress suggests that the multimodality of the fitness landscape is of importance in determining the relative performance of recombination and mutation. This observation will be explored further in Chap. 14.

There are a variety of directions worth exploring. The first would be to expand the mathematical techniques to include variances as well as expectations. For example, the variance of the mean first passage times is an important measure that may also be derivable from these models. Second, the model itself can be generalized to include other operators (e.g., uniform recombination) and other EA features such as rank selection, elitism, and so on. Finally, an interesting future possibility is to create Markov models of other search algorithms, and then allow a meta-search algorithm to find those problems that are easy for one search algorithm and hard for another (De Jong et al. 1994). As a simple example of this, suppose the expected waiting time of an arbitrary function was computed for an EA with recombination and an

EA without recombination. Then a meta-search algorithm (possibly an EA) could search the space of functions in an attempt to maximize the difference in the expected waiting times. Such automated techniques could also be used with the search algorithms themselves (rather than Markov chain models of the algorithms), yielding an automatic technique for discovering those classes of problems that hard for one algorithm and easy for another.

As pointed out earlier, one primary concern is the general applicability of such Markov chain approaches, given the huge number of states that arise with even small problems and population sizes. Chapter 11 provided an example of a simpler model that could be "aggregated" naturally for a particular problem class. However, this is much more difficult to accomplish on the more complex model described in this chapter. The question, then, is whether such complex models can be automatically aggregated. The answer is *yes*. Chapter 13 addresses this issue in great detail, and proposes a novel aggregation algorithm that can automatically compress a Markov chain of N states into an aggregated Markov chain with far fewer states, without introducing significant numerical error.

13. An Aggregation Algorithm for Markov Chains

13.1 Introduction

Chapter 12 uses a Markov chain model (Nix and Vose 1992) of a complete finite-population EA with selection, mutation, and recombination. Each state of the Markov model is a particular population of the EA. If there are N states, then the Markov chain model is defined by an $N \times N$ matrix Q called the "one-step probability transition matrix," where $Q(i,j)$ is the probability of going from state i to state j in one step. The n-step (transient) behavior of the system is described by the nth power of Q, Q^n. For EAs, the number of states grows enormously as the population size (or string length) increases (e.g., see Table 12.1), which can make the models computationally intensive. Our motivation for examining the Markov chain models was to explore the differences between mutation and recombination in EAs. However, due to the large size of the models, this chapter makes an excursion and introduces a novel technique for simplifying Markov models, in order to automatically reduce the number of states in the model.

Previous methods for reducing the number of states (referred to as *compression*, *aggregation*, or *lumping* methods) have focused on techniques that provide good estimations of the steady-state behavior of the Markov model (e.g., see Stewart 1994). The focus of this chapter, however, is on *transient* behavior, and the goal is to produce an algorithm for aggregating Q matrices in a way that yields good estimates of the transient behavior of the Markov model. The algorithm described in this chapter aggregates a Q matrix into a smaller Q matrix with fewer states. In general, the aggregation will not be without error, so the goal is to provide an algorithm that aggregates the original Q matrix without significant error. Although computing an aggregated matrix might take some time, the savings resulting from using this aggregated matrix in all subsequent computations can more than offset the aggregation time.

The organization of this chapter is as follows. Section 13.2 introduces the aggregation algorithm, which aggregates pairs of states by taking a *weighted* average of the row entries of Q for those two states, followed by summing the two columns of Q associated with those two states. Section 13.2 also introduces the important concepts of row and column *equivalence*, which are important for identifying pairs of states that can be aggregated with no error. Section 13.3 provides mathematical justification for taking the weighted average of row entries and shows that the weights are simply column sums of probability mass. Section 13.4 proves that pairs of states that are row or column equivalent lead to perfect aggregation. Section 13.5 introduces an analysis of error and uses this to define a metric for row and column *similarity* which can be used to find pairs of states that yield almost perfect aggregation. Later sections illustrate the utility of the aggregation algorithm through experiments.

13.2 The Aggregation Algorithm at a High Level

The entries in the Q matrix, $p_{i,j} \equiv Q(i,j)$, represent the conditional probability that the system will transition to state j in one step, given that it currently is in state i.[1] Now suppose that states i and j have been chosen for aggregation. The new aggregated state is referred to as state $\{i \vee j\}$. Aggregating states i and j together means that the combined state represents being in either state i or state j. Since this is a disjunctive situation, the probability of transition from state k *into* the aggregated state is simply the sum $p_{k,\{i \vee j\}} = p_{k,i} + p_{k,j}$. Stated another way, part of the aggregation algorithm is to sum columns of probability numbers in Q.

However, in general, transitions *from* an aggregated state are more complicated to compute. Clearly, the probability $p_{\{i \vee j\},k}$ of transitioning from the aggregated state to some other state must lie somewhere between $p_{i,k}$ and $p_{j,k}$, depending on how much time is spent in states i and j. Thus a weighted average of row entries in Q appears to be called for, where the weights reflect the amount of time spent in states i and j. Precisely how to do this weighted average is investigated in Sect. 13.3.

The algorithm for aggregating two states i and j together is as follows:[2]

Aggregate-states(i,j)
- (a) Compute a weighted average of the ith and jth rows. Place the results in rows i and j.
- (b) Sum the ith and jth columns. Place the results in column i. Remove row j and column j.

The aggregation algorithm has two steps. It takes as input a matrix Q_u (an unaggregated Q matrix). Step (a) averages the row entries, producing an intermediate row-averaged matrix Q_r. Step (b) sums column entries to produce the final aggregated (compressed) matrix Q_c. Step (a) is the sole

[1] The notation $p_{i,j}{}^{(n)} \equiv Q^n(i,j)$ denotes the entries of the n-step probability transition matrix Q^n.

[2] The algorithm is written this way because it makes it amenable to mathematical analysis.

source of error, since in general it is difficult to estimate the amount of time spent in states i and j.

Now that the aggregation algorithm has been outlined, it is important to define what is meant by "perfect" aggregation. As mentioned before, analysis of n-step transition probabilities (i.e., *transient* behavior of the Markov chain) can be realized by computing Q^n. For large Q matrices this is computationally expensive. It would be less expensive to aggregate Q and to then raise it to the nth power. If the aggregation algorithm has worked well then the nth power of the aggregated matrix Q_c should be (nearly) identical to aggregating the nth power of the unaggregated matrix Q_u. In other words, perfect aggregation has occurred if $({Q_u}^n)_c = {Q_c}^n$.

It turns out that there are two situations under which perfect aggregation can be obtained. The first situation is referred to as "row equivalence," in which the two states i and j have identical rows (i.e., $\forall k \; p_{i,k} = p_{j,k}$). In this case the weighted averaging cannot produce any error, since the weights will be irrelevant. The second situation is referred to as "column equivalence," in which state i has column entries that are a real multiple q of the column entries for state j (i.e., $\forall k \; p_{k,i} = qp_{k,j}$). The intuition here is that when this situation occurs, the ratio of time spent in state i to state j is precisely q. The details of this can be found in Sect. 13.4.

However, for arbitrary matrices, aggregating an arbitrarily chosen pair of states will not necessarily lead to good results. Thus, the goal is to identify pairs of states i and j upon which the above aggregation algorithm will work well. It turns out that pairs of states that are row or column *similar* are good candidates for aggregation. The justification for these measures will be provided in Sect. 13.5.

At a high level, of course, this simple aggregation algorithm must be repeated for many pairs of states, if one wants to dramatically reduce the size of a Q matrix. The high level aggregation algorithm is simply:

Aggregate()
  Repeat as long as possible
    (i) Find the pair of states i and j most similar to each other.
    (ii) Aggregate-states(i,j).

13.3 The Aggregation Algorithm in More Detail

In the previous section the aggregation algorithm was described in two steps. Step (a) is where error can occur and care must be taken to mathematically justify the weighted averaging of rows. This can be done by attempting to force $({Q_u}^2)_c$ to be as similar as possible to ${Q_c}^2$ (later sections will generalize this to higher powers). This is mathematically difficult, but fortunately it suffices to force ${Q_u}^2$ to be as similar as possible to $Q_u Q_r$, which is much simpler and focuses on the row-averaged matrix Q_r explicitly. The intuition behind this is that if aggregation is done correctly, passage through the new aggregated state should affect the 2-step transition probabilities as little as possible.[3] This will be shown with a 4×4 Q matrix, and then generalized to an arbitrary $N \times N$ matrix. The result will be the weighted row-averaging procedure outlined earlier. This particular presentation has been motivated by a concern for comprehension and hence is not completely formal. A completely formal presentation is in the Appendix.

13.3.1 Weighted Averaging with a 4 × 4 Matrix

Consider a general unaggregated 4×4 matrix Q_u for a Markov chain model of 4 states, as well as the general intermediate matrix Q_r:

$$Q_\mathrm{u} = \begin{bmatrix} p_{1,1} & p_{1,2} & p_{1,3} & p_{1,4} \\ p_{2,1} & p_{2,2} & p_{2,3} & p_{2,4} \\ p_{3,1} & p_{3,2} & p_{3,3} & p_{3,4} \\ p_{4,1} & p_{4,2} & p_{4,3} & p_{4,4} \end{bmatrix} \qquad Q_\mathrm{r} = \begin{bmatrix} r_{1,1} & r_{1,2} & r_{1,3} & r_{1,4} \\ r_{2,1} & r_{2,2} & r_{2,3} & r_{2,4} \\ r_{3,1} & r_{3,2} & r_{3,3} & r_{3,4} \\ r_{4,1} & r_{4,2} & r_{4,3} & r_{4,4} \end{bmatrix}$$

The notation $r_{i,j} \equiv Q_\mathrm{r}(i,j)$ is used to prevent confusion with the $p_{i,j}$ in Q_u. Without loss of generality the goal will be to aggregate the third and fourth states (rows and columns) of this matrix. Since the third and fourth states are being aggregated, rows 1 and 2 of Q_r must be the same as Q_u (i.e., averaging rows 3 and 4 will not affect rows 1 and 2). Denoting $\{3 \vee 4\}$ to be the aggregated state, the intermediate matrix is:

$$Q_\mathrm{r} = \begin{bmatrix} p_{1,1} & p_{1,2} & p_{1,3} & p_{1,4} \\ p_{2,1} & p_{2,2} & p_{2,3} & p_{2,4} \\ r_{\{3\vee 4\},1} & r_{\{3\vee 4\},2} & r_{\{3\vee 4\},3} & r_{\{3\vee 4\},4} \\ r_{\{3\vee 4\},1} & r_{\{3\vee 4\},2} & r_{\{3\vee 4\},3} & r_{\{3\vee 4\},4} \end{bmatrix}$$

The $r_{\{3\vee 4\},k}$ represent the weighted average of rows 3 and 4 of Q_u. Recall that step (a) of Aggregate-states(3,4) will place that average in both rows 3 and 4, which is why rows 3 and 4 of Q_r are the same. The trick now is to determine what $r_{\{3\vee 4\},1}$, $r_{\{3\vee 4\},2}$, $r_{\{3\vee 4\},3}$, and $r_{\{3\vee 4\},4}$ should be in order to produce a reasonable aggregation. This is done by considering ${Q_\mathrm{u}}^2$ and $Q_\mathrm{u}Q_\mathrm{r}$.

$${Q_\mathrm{u}}^2 = \begin{bmatrix} {p_{1,1}}^{(2)} & {p_{1,2}}^{(2)} & {p_{1,3}}^{(2)} & {p_{1,4}}^{(2)} \\ {p_{2,1}}^{(2)} & {p_{2,2}}^{(2)} & {p_{2,3}}^{(2)} & {p_{2,4}}^{(2)} \\ {p_{3,1}}^{(2)} & {p_{3,2}}^{(2)} & {p_{3,3}}^{(2)} & {p_{3,4}}^{(2)} \\ {p_{4,1}}^{(2)} & {p_{4,2}}^{(2)} & {p_{4,3}}^{(2)} & {p_{4,4}}^{(2)} \end{bmatrix}$$

[3] More formally, it can be shown that if ${Q_\mathrm{u}}^2 = Q_\mathrm{u}Q_\mathrm{r}$ then $({Q_\mathrm{u}}^2)_\mathrm{c} = {Q_\mathrm{c}}^2$ for row- or column-equivalent situations. See Sect. 13.4.

$$Q_{\mathrm{u}}Q_{\mathrm{r}} = \begin{bmatrix} a_{1,1}{}^{(2)} & a_{1,2}{}^{(2)} & a_{1,3}{}^{(2)} & a_{1,4}{}^{(2)} \\ a_{2,1}{}^{(2)} & a_{2,2}{}^{(2)} & a_{2,3}{}^{(2)} & a_{2,4}{}^{(2)} \\ a_{3,1}{}^{(2)} & a_{3,2}{}^{(2)} & a_{3,3}{}^{(2)} & a_{3,4}{}^{(2)} \\ a_{4,1}{}^{(2)} & a_{4,2}{}^{(2)} & a_{4,3}{}^{(2)} & a_{4,4}{}^{(2)} \end{bmatrix}$$

The notation $a_{i,j}{}^{(2)}$ is used to prevent confusion with the $p_{i,j}{}^{(2)}$ in $Q_{\mathrm{u}}{}^2$. Since the goal is to have $Q_{\mathrm{u}}{}^2 = Q_{\mathrm{u}}Q_{\mathrm{r}}$, it is necessary to have $p_{i,j}{}^{(2)}$ be as similar as possible to $a_{i,j}{}^{(2)}$. The $p_{i,j}{}^{(2)}$ values can be computed using $p_{i,j}$ values, while the $a_{i,j}{}^{(2)}$ values require the unknowns $r_{\{3\vee4\},1}$, $r_{\{3\vee4\},2}$, $r_{\{3\vee4\},3}$, and $r_{\{3\vee4\},4}$.

For example, $p_{1,1}{}^{(2)}$ can be computed by multiplying Q_{u} by itself:

$$p_{1,1}{}^{(2)} = p_{1,1}p_{1,1} + p_{1,2}p_{2,1} + p_{1,3}p_{3,1} + p_{1,4}p_{4,1}$$

However, $a_{1,1}{}^{(2)}$ is computed by multiplying Q_{u} and Q_{r}:

$$a_{1,1}{}^{(2)} = p_{1,1}p_{1,1} + p_{1,2}p_{2,1} + (p_{1,3} + p_{1,4})r_{\{3\vee4\},1}$$

In the ideal situation we would like both of these to be equal. This implies that:

$$r_{\{3\vee4\},1} = \frac{p_{1,3}p_{3,1} + p_{1,4}p_{4,1}}{p_{1,3} + p_{1,4}}$$

But we can write another formula for $r_{\{3\vee4\},1}$ by considering $p_{2,1}{}^{(2)}$ and $a_{2,1}{}^{(2)}$:

$$p_{2,1}{}^{(2)} = p_{2,1}p_{1,1} + p_{2,2}p_{2,1} + p_{2,3}p_{3,1} + p_{2,4}p_{4,1}$$
$$a_{2,1}{}^{(2)} = p_{2,1}p_{1,1} + p_{2,2}p_{2,1} + (p_{2,3} + p_{2,4})r_{\{3\vee4\},1}$$

Again, we would like both of these to be equal. This implies that:

$$r_{\{3\vee4\},1} = \frac{p_{2,3}p_{3,1} + p_{2,4}p_{4,1}}{p_{2,3} + p_{2,4}}$$

Similarly, consideration of $p_{3,1}{}^{(2)}$ and $a_{3,1}{}^{(2)}$ yields:

$$r_{\{3\vee4\},1} = \frac{p_{3,3}p_{3,1} + p_{3,4}p_{4,1}}{p_{3,3} + p_{3,4}}$$

while consideration of $p_{4,1}{}^{(2)}$ and $a_{4,1}{}^{(2)}$ yields:

$$r_{\{3\vee4\},1} = \frac{p_{4,3}p_{3,1} + p_{4,4}p_{4,1}}{p_{4,3} + p_{4,4}}$$

What has happened here is that the four elements in the first column of $Q_{\mathrm{u}}Q_{\mathrm{r}}$ lead to four expressions for $r_{\{3\vee4\},1}$. In general, all four expressions for

$r_{\{3\vee 4\},1}$ cannot hold simultaneously (although we will investigate conditions under which they will hold later). The best estimate is to take a weighted average of the four expressions for $r_{\{3\vee 4\},1}$ (this is related to the concept of "averaging" probabilities – see Appendix for more details). This yields:

$$r_{\{3\vee 4\},1} = \frac{(p_{1,3}+p_{2,3}+p_{3,3}+p_{4,3})p_{3,1}+(p_{1,4}+p_{2,4}+p_{3,4}+p_{4,4})p_{4,1}}{(p_{1,3}+p_{2,3}+p_{3,3}+p_{4,3})+(p_{1,4}+p_{2,4}+p_{3,4}+p_{4,4})}$$

Note how the final expression for $r_{\{3\vee 4\},1}$ is a weighted average of the row entries $p_{3,1}$ and $p_{4,1}$, where the weights are column sums for columns 3 and 4. In general the elements of $Q_{\mathrm{u}}Q_{\mathrm{r}}$ in the kth column will constrain $r_{\{3\vee 4\},k}$:

$$r_{\{3\vee 4\},k} = \frac{(p_{1,3}+p_{2,3}+p_{3,3}+p_{4,3})p_{3,k}+(p_{1,4}+p_{2,4}+p_{3,4}+p_{4,4})p_{4,k}}{(p_{1,3}+p_{2,3}+p_{3,3}+p_{4,3})+(p_{1,4}+p_{2,4}+p_{3,4}+p_{4,4})}$$

Once again, note how the expression for $r_{\{3\vee 4\},k}$ is a weighted average of the row entries $p_{3,k}$ and $p_{4,k}$, where the weights are column sums for columns 3 and 4.

13.3.2 Weighted Averaging with an $N \times N$ Matrix

The previous results for a 4×4 matrix can be extended to an $N \times N$ matrix. Without loss of generality aggregate states $N-1$ and N. Then the N elements of column k yield N expressions for each $r_{\{N-1\vee N\},k}$. The best estimate is (see Appendix for details):

$$r_{\{N-1\vee N\},k} = \frac{(p_{1,N-1}+\ldots+p_{N,N-1})p_{N-1,k}+(p_{1,N}+\ldots+p_{N,N})p_{N,k}}{(p_{1,N-1}+\ldots+p_{N,N-1})+(p_{1,N}+\ldots+p_{N,N})}$$

Note again how the weights are column sums for columns $N-1$ and N. Generalizing this to aggregating two arbitrary states i and j yields:

$$r_{\{i\vee j\},k} = \frac{(\sum_l p_{l,i})p_{i,k}+(\sum_l p_{l,j})p_{j,k}}{\sum_l p_{l,i}+\sum_l p_{l,j}}$$

or:

$$r_{\{i\vee j\},k} = \frac{m_i p_{i,k}+m_j p_{j,k}}{m_i+m_j} \tag{13.1}$$

where m_i and m_j are the sums of the probability mass in columns i and j of Q_{u}.

Equation 13.1 indicates how to compute the $r_{\{i\vee j\},k}$ entries in Q_{r}. Note how they are computed using the weighted average of the row entries in rows i and j. The weights are simply the column sums. This justifies the row-averaging component of the aggregation algorithm described in the previous section. Intuitively stated, the column mass for columns i and j provide good estimates of the relative amount of time spent in states i and j. The estimates are used as weights to average the transitions from i to state k and from j to k, producing the probability of transition from the combined state $\{i\vee j\}$ to k.

13.3.3 Mathematical Restatement of the Aggregation Algorithm

Now that the weighted averaging of rows i and j has been explained, it is only necessary to sum columns i and j in order to complete the aggregation algorithm. The whole algorithm can be expressed simply as follows. Assume that two states have been chosen for aggregation. Let S denote the set of all N states, and let the nonempty sets $S_1, ..., S_{N-1}$ partition S such that one S_i contains the two chosen states, while each other S_i is composed of exactly one state. Let m_i denote the column mass of state i. Then the aggregated (compressed) matrix Q_c is:

$$Q_c(x, y) = \frac{1}{\sum_{i \in S_x} m_i} \sum_{i \in S_x} \left[m_i \sum_{j \in S_y} p_{i,j} \right] \tag{13.2}$$

This corresponds to taking a weighted average of the two rows corresponding to the two chosen states, while summing the two corresponding columns. The other entries in the Q matrix remain unchanged. Consider an example (where $N = 3$) in which states 2 and 3 are aggregated. In that case $S_1 = \{1\}$ and $S_2 = \{2, 3\}$. Q_c is described by:

$$\begin{aligned}
Q_c(1,1) &= p_{1,1} \\
Q_c(1,2) &= p_{1,2} + p_{1,3} \\
Q_c(2,1) &= \frac{1}{m_2 + m_3}[m_2 p_{2,1} + m_3 p_{3,1}] \\
Q_c(2,2) &= \frac{1}{m_2 + m_3}[m_2(p_{2,2} + p_{2,3}) + m_3(p_{3,2} + p_{3,3})]
\end{aligned}$$

Applying this to the following column-equivalent matrix Q_u produces perfect results $(({Q_u}^2)_c = {Q_c}^2)$:

$$Q_u = \begin{bmatrix} .7 & .1 & .2 \\ .4 & .2 & .4 \\ .1 & .3 & .6 \end{bmatrix} \Rightarrow {Q_u}^2 = \begin{bmatrix} .55 & .15 & .30 \\ .40 & .20 & .40 \\ .25 & .25 & .50 \end{bmatrix} \Rightarrow$$

$$({Q_u}^2)_c = \begin{bmatrix} .55 & .45 \\ .30 & .70 \end{bmatrix}$$

$$Q_c = \begin{bmatrix} .7 & .3 \\ .2 & .8 \end{bmatrix} \Rightarrow {Q_c}^2 = \begin{bmatrix} .55 & .45 \\ .30 & .70 \end{bmatrix}$$

In summary, this section has justified the use of column mass as weights in the row-averaging portion of the aggregation algorithm. The whole aggregation algorithm is stated succinctly as a mathematical function, which can aggregate any arbitrary pair of states. However, as stated earlier, aggregation of arbitrary pairs of states need not lead to good aggregation. The goal, then, is to identify good pairs of states. This is investigated in the next section, and relies upon the concepts of row and column equivalence.

13.4 Special Cases in Which Aggregation Is Perfect

If aggregation is working well, then the aggregated version of ${Q_\mathrm{u}}^n$ should be (nearly) identical to ${Q_\mathrm{c}}^n$. As suggested in Sect. 13.2, there are two situations under which perfect aggregation will occur. The first situation is when two states are row equivalent. The intuition here is that the row average of two identical rows will not involve any error, and thus the aggregation will be perfect. The second situation is when two states are column equivalent. The intuition for this situation is that if the column $\mathbf{b}_i$ is equal to $q\mathbf{b}_j$, then the ratio of time spent in state i to state j is exactly q. Under these circumstances the weighted row average will also produce no error.

This section will prove that $({Q_\mathrm{u}}^n)_\mathrm{c} = {Q_\mathrm{c}}^n$ when the two states being aggregated are either row equivalent or column equivalent. This will hold for any n and for any Q_u matrix of size $N \times N$. The method of proof will be to treat the aggregation algorithm as a linear transformation f, and then to show that $f({Q_\mathrm{u}}^n) = (f(Q_\mathrm{u}))^n$, where $f(Q_\mathrm{u}) = Q_\mathrm{c}$.

13.4.1 Row Equivalence and the Aggregation Algorithm

This subsection will prove that when two states are row equivalent, aggregation of those states can be described by a linear transformation (matrix multiplication). The aggregation algorithm aggregates an $N \times N$ matrix Q_u to an $(N-1) \times (N-1)$ matrix Q_c. However, for the sake of mathematical convenience all of the matrix transformations will be with $N \times N$ matrices. Without loss of generality it is assumed that states $N-1$ and N are being aggregated. When it comes time to expressing the final aggregation, the Nth row and column will simply be ignored, producing the $(N-1) \times (N-1)$ aggregated matrix. The "•" denotes an unimportant entry.

Assume that states $N-1$ and N are row equivalent. Thus $\forall k \; p_{N-1,k} = p_{N,k}$. Using Eq. 13.1 to compute the row averages yields:

$$r_{\{N-1 \vee N\},k} = \frac{m_{N-1}p_{N-1,k} + m_N p_{N,k}}{m_{N-1} + m_N} = \frac{p_{N-1,k}(m_{N-1} + m_N)}{m_{N-1} + m_N} = p_{N-1,k}$$

and the aggregated matrix should have the form:

$$Q_\mathrm{c} = \begin{bmatrix} p_{1,1} & \cdots & p_{1,N-2} & p_{1,N-1} + p_{1,N} \\ p_{2,1} & \cdots & p_{2,N-2} & p_{2,N-1} + p_{2,N} \\ \vdots & & \vdots & \vdots \\ p_{N-1,1} & \cdots & p_{N-1,N-2} & p_{N-1,N-1} + p_{N-1,N} \end{bmatrix}$$

Theorem 13.4.1. *If states N and $N-1$ in Q_u are row equivalent then $Q_\mathrm{c} = TQ_\mathrm{u}T$ and $TT = I$, where*

$$T = \left[\begin{array}{c|cc} I & \multicolumn{2}{c}{0} \\ \hline 0 & 1 & 0 \\ & 1 & -1 \end{array}\right]$$

Proof. $Q_c = TQ_uT$ can be expressed as follows:

$$Q_c = T \begin{bmatrix} p_{1,1} & \cdots & p_{1,N-2} & p_{1,N-1} & p_{1,N} \\ p_{2,1} & \cdots & p_{2,N-2} & p_{2,N-1} & p_{2,N} \\ \vdots & & \vdots & \vdots & \vdots \\ p_{N-1,1} & \cdots & p_{N-1,N-2} & p_{N-1,N-1} & p_{N-1,N} \\ p_{N,1} & \cdots & p_{N,N-2} & p_{N,N-1} & p_{N,N} \end{bmatrix} \left[\begin{array}{c|cc} I & \multicolumn{2}{c}{0} \\ \hline 0 & 1 & 0 \\ & 1 & -1 \end{array}\right] =$$

$$\left[\begin{array}{c|cc} I & \multicolumn{2}{c}{0} \\ \hline 0 & 1 & 0 \\ & 1 & -1 \end{array}\right] \begin{bmatrix} p_{1,1} & \cdots & p_{1,N-2} & p_{1,N-1} + p_{1,N} & \bullet \\ p_{2,1} & \cdots & p_{2,N-2} & p_{2,N-1} + p_{2,N} & \bullet \\ \vdots & & \vdots & \vdots & \vdots \\ p_{N-1,1} & \cdots & p_{N-1,N-2} & p_{N-1,N-1} + p_{N-1,N} & \bullet \\ p_{N-1,1} & \cdots & p_{N-1,N-2} & p_{N-1,N-1} + p_{N-1,N} & \bullet \end{bmatrix} =$$

$$\begin{bmatrix} p_{1,1} & \cdots & p_{1,N-2} & p_{1,N-1} + p_{1,N} & \bullet \\ p_{2,1} & \cdots & p_{2,N-2} & p_{2,N-1} + p_{2,N} & \bullet \\ \vdots & & \vdots & \vdots & \vdots \\ p_{N-1,1} & \cdots & p_{N-1,N-2} & p_{N-1,N-1} + p_{N-1,N} & \bullet \\ \bullet & \cdots & \bullet & \bullet & \bullet \end{bmatrix}$$

□

This is precisely what Q_c should be. Thus the aggregation of two row-equivalent states can be expressed simply as TQ_uT. The first T performs row averaging (which is trivial) and the second T performs column summing. The reader will also note that some elements of T do not appear to be important for the derivation that $Q_c = TQ_uT$. This is true, however, the purpose of these elements is to ensure that $TT = I$, since this fact will also be used to help prove that $(Q_u{}^n)_c = Q_c{}^n$.

$$TT = \left[\begin{array}{c|cc} I & \multicolumn{2}{c}{0} \\ \hline 0 & 1 & 0 \\ & 1 & -1 \end{array}\right] \left[\begin{array}{c|cc} I & \multicolumn{2}{c}{0} \\ \hline 0 & 1 & 0 \\ & 1 & -1 \end{array}\right] = I$$

13.4.2 Column Equivalence and the Aggregation Algorithm

This subsection will prove that when two states are column equivalent, aggregation of those states can be described by a linear transformation. Assume without loss of generality that states $N-1$ and N are column equivalent. Thus $\forall k \; p_{k,N-1} = q p_{k,N}$, and $m_{N-1} = q m_N$. Using Eq. 13.1 to compute the row averages yields:

$$r_{\{N-1 \vee N\},k} = \frac{m_{N-1}p_{N-1,k} + m_N p_{N,k}}{m_{N-1} + m_N} = \frac{q p_{N-1,k} + p_{N,k}}{q+1}$$

and the aggregated matrix should have the form:

$$Q_c = \begin{bmatrix} p_{1,1} & \cdots & p_{1,N-2} & p_{1,N-1} + p_{1,N} \\ p_{2,1} & \cdots & p_{2,N-2} & p_{2,N-1} + p_{2,N} \\ \vdots & & \vdots & \vdots \\ \frac{qp_{N-1,1}+p_{N,1}}{q+1} & \cdots & \frac{qp_{N-1,N-2}+p_{N,N-2}}{q+1} & \frac{qp_{N-1,N-1}+p_{N,N-1}+qp_{N-1,N}+p_{N,N}}{q+1} \end{bmatrix}$$

Theorem 13.4.2. *If states N and $N-1$ in Q_u are column equivalent then $Q_c = XQ_uY$ and $YX = I$ where*

$$Y = \left[\begin{array}{c|cc} I & \multicolumn{2}{c}{0} \\ \hline 0 & 1 & \frac{1}{q} \\ & 1 & -1 \end{array}\right] \quad X = \left[\begin{array}{c|cc} I & \multicolumn{2}{c}{0} \\ \hline 0 & \frac{q}{q+1} & \frac{1}{q+1} \\ & \frac{q}{q+1} & -\frac{q}{q+1} \end{array}\right]$$

Proof. $Q_c = XQ_uY$ can be expressed as follows:

$$Q_c = X \begin{bmatrix} p_{1,1} & \cdots & p_{1,N-2} & p_{1,N-1} & p_{1,N} \\ p_{2,1} & \cdots & p_{2,N-2} & p_{2,N-1} & p_{2,N} \\ \vdots & & \vdots & \vdots & \vdots \\ p_{N-1,1} & \cdots & p_{N-1,N-2} & p_{N-1,N-1} & p_{N-1,N} \\ p_{N,1} & \cdots & p_{N,N-2} & p_{N,N-1} & p_{N,N} \end{bmatrix} \left[\begin{array}{c|cc} I & \multicolumn{2}{c}{0} \\ \hline 0 & 1 & \frac{1}{q} \\ & 1 & -1 \end{array}\right] =$$

$$X \begin{bmatrix} p_{1,1} & \cdots & p_{1,N-2} & p_{1,N-1} + p_{1,N} \\ p_{2,1} & \cdots & p_{2,N-2} & p_{2,N-1} + p_{2,N} \\ \vdots & & \vdots & \vdots \\ p_{N-1,1} & \cdots & p_{N-1,N-2} & p_{N-1,N-1} + p_{N-1,N} \\ p_{N,1} & \cdots & p_{N,N-2} & p_{N,N-1} + p_{N,N} \end{bmatrix} =$$

$$\begin{bmatrix} p_{1,1} & \cdots & p_{1,N-2} & p_{1,N-1} + p_{1,N} \\ p_{2,1} & \cdots & p_{2,N-2} & p_{2,N-1} + p_{2,N} \\ \vdots & & \vdots & \vdots \\ \frac{qp_{N-1,1}+p_{N,1}}{q+1} & \cdots & \frac{qp_{N-1,N-2}+p_{N,N-2}}{q+1} & \frac{qp_{N-1,N-1}+p_{N,N-1}+qp_{N-1,N}+p_{N,N}}{q+1} \end{bmatrix}$$

□

This is precisely what Q_c should be.[4] Thus the aggregation of a pair of column-equivalent states can be expressed simply as $XQ_\mathrm{u}Y$. X performs row averaging and Y performs column summing. The reader will note that some elements of X and Y are not important for the derivation that $Q_\mathrm{c} = XQ_\mathrm{u}Y$ (e.g., T could be used instead of Y). This is true, however, the purpose of these elements is to ensure that $YX = I$, since this fact will be used to help prove that $(Q_\mathrm{u}{}^n)_\mathrm{c} = Q_\mathrm{c}{}^n$ at the end of this section.

$$YX = \left[\begin{array}{c|cc} I & \multicolumn{2}{c}{0} \\ \hline 0 & 1 & \frac{1}{q} \\ & 1 & -1 \end{array}\right] \left[\begin{array}{c|cc} I & \multicolumn{2}{c}{0} \\ \hline 0 & \frac{q}{q+1} & \frac{1}{q+1} \\ & \frac{q}{q+1} & -\frac{q}{q+1} \end{array}\right] = I$$

13.4.3 Some Necessary Lemmas

Before proving that $(Q_\mathrm{u}{}^n)_\mathrm{c} = Q_\mathrm{c}{}^n$ for row- or column-equivalent states, it is necessary to prove some simple lemmas. The idea is to show that if Q_u is row or column equivalent, so is $Q_\mathrm{u}{}^n$. This will allow the previous linear transformations to be applied to $Q_\mathrm{u}{}^n$ as well as Q_u.

Let square matrices A and B be defined as matrices of row and column vectors respectively:

$$A = \begin{bmatrix} a_{1,1} & \cdots & a_{1,N} \\ \vdots & & \vdots \\ a_{N,1} & \cdots & a_{N,N} \end{bmatrix} = \begin{bmatrix} \mathbf{a}_1 \\ \vdots \\ \mathbf{a}_N \end{bmatrix}$$

$$B = \begin{bmatrix} b_{1,1} & \cdots & b_{1,N} \\ \vdots & & \vdots \\ b_{N,1} & \cdots & b_{N,N} \end{bmatrix} = \begin{bmatrix} \mathbf{b}_1 & \cdots & \mathbf{b}_N \end{bmatrix}$$

Then the matrix product AB can be represented using dot product notation:

$$AB = \begin{bmatrix} \mathbf{a}_1 \cdot \mathbf{b}_1 & \cdots & \mathbf{a}_1 \cdot \mathbf{b}_N \\ \vdots & & \vdots \\ \mathbf{a}_N \cdot \mathbf{b}_1 & \cdots & \mathbf{a}_N \cdot \mathbf{b}_N \end{bmatrix}$$

Lemma 13.4.1. *Row equivalence is invariant under post-multiplication.*

Proof. Suppose states i and j of A are row equivalent ($\mathbf{a}_i = \mathbf{a}_j$). Then $\forall k\ \mathbf{a}_i \cdot \mathbf{b}_k = \mathbf{a}_j \cdot \mathbf{b}_k$. So, states i and j in AB must be row equivalent. □

Lemma 13.4.2. *Column equivalence is invariant under pre-multiplication.*

[4] For this proof we have omitted the irrelevant "•" entries.

Proof. Suppose states i and j of B are column equivalent ($\mathbf{b}_i = q\mathbf{b}_j$). Then $\forall k \ \mathbf{a}_k \cdot \mathbf{b}_i = q\mathbf{a}_k \cdot \mathbf{b}_j$. So, states i and j in AB must be column equivalent. □

Lemma 13.4.3. *Row and column equivalence are invariant under raising to a power.*

Proof. $Q^n = QQ^{n-1}$. Thus, if states i and j are row equivalent in Q, they are row equivalent in Q^n by Lemma 13.4.1. Similarly, $Q^n = Q^{n-1}Q$. Thus, if states i and j are column equivalent in Q, they are column equivalent in Q^n by Lemma 13.4.2. □

Lemma 13.4.3 indicates that the previous linear transformations can be applied to ${Q_\mathrm{u}}^n$ to produce $({Q_\mathrm{u}}^n)_\mathrm{c}$ when two states in Q_u are row or column equivalent.

13.4.4 Theorems for Perfect Aggregation

Given the previous theorems concerning the linear transformations and Lemma 13.4.3, it is now possible to state and prove the theorems for perfect aggregation. The Q matrix can be considered to be Q_u in these theorems.

Theorem 13.4.3. *If Q is row equivalent, then $Q^n = Q{Q_\mathrm{r}}^{n-1}$ implies $(Q^n)_\mathrm{r} = {Q_\mathrm{r}}^n$, and $(Q^n)_\mathrm{c} = {Q_\mathrm{c}}^n$*

Proof. If Q is row equivalent, then so is Q^n by Lemma 13.4.3. If $Q^n = Q{Q_\mathrm{r}}^{n-1}$ then $(Q^n)_\mathrm{r} = (Q{Q_\mathrm{r}}^{n-1})_\mathrm{r} = TQ{Q_\mathrm{r}}^{n-1} = {Q_\mathrm{r}}^n$, and $(Q^n)_\mathrm{c} = (Q{Q_\mathrm{r}}^{n-1})_\mathrm{c} = T(Q{Q_\mathrm{r}}^{n-1})T = TQ{Q_\mathrm{c}}^{n-1} = TQTT{Q_\mathrm{c}}^{n-1} = Q_\mathrm{c}T{Q_\mathrm{c}}^{n-1} = {Q_\mathrm{c}}^n$. □

Theorem 13.4.4. *If Q is column equivalent, then $Q^n = Q{Q_\mathrm{r}}^{n-1}$ implies $(Q^n)_\mathrm{r} = {Q_\mathrm{r}}^n$, and $(Q^n)_\mathrm{c} = {Q_\mathrm{c}}^n$*

Proof. If Q is column equivalent, then so is Q^n by Lemma 13.4.3. If $Q^n = Q{Q_\mathrm{r}}^{n-1}$ then $(Q^n)_\mathrm{r} = (Q{Q_\mathrm{r}}^{n-1})_\mathrm{r} = XQ{Q_\mathrm{r}}^{n-1} = {Q_\mathrm{r}}^n$, and $(Q^n)_\mathrm{c} = (Q{Q_\mathrm{r}}^{n-1})_\mathrm{c} = X(Q{Q_\mathrm{r}}^{n-1})Y = XQ{Q_\mathrm{c}}^{n-1} = XQTT{Q_\mathrm{c}}^{n-1} = Q_\mathrm{c}T{Q_\mathrm{c}}^{n-1} = {Q_\mathrm{c}}^n$. □

These two theorems illustrate the validity of trying to force ${Q_\mathrm{u}}^2$ to be as similar as possible to $Q_\mathrm{u}Q_\mathrm{r}$ in Sect. 13.3.

Theorem 13.4.5. *If Q is row equivalent, then $(Q^n)_\mathrm{c} = {Q_\mathrm{c}}^n$.*

Proof. If Q is row equivalent, then so is Q^n by Lemma 13.4.3. Then $(Q^n)_\mathrm{c} = TQ^nT = TQ \cdots QT$. Since $TT = I$, it is simple to show that $(Q^n)_\mathrm{c} = TQTTQ \cdots QTTQT = {Q_\mathrm{c}}^n$. □

Theorem 13.4.6. *If Q is column equivalent, then $(Q^n)_\mathrm{c} = {Q_\mathrm{c}}^n$.*

Proof. If Q is column equivalent, then Q^n is by Lemma 13.4.3. Then $(Q^n)_\mathrm{c} = XQ^nY = XQ \cdots QY$. Since $YX = I$, then $(Q^n)_\mathrm{c} = XQYXQ \cdots QYXQY = {Q_\mathrm{c}}^n$. □

These theorems hold for all n and for all row- or column-equivalent $N \times N$ Q matrices, and highlight the importance of row and column equivalence. If two states are row or column equivalent, then aggregation of those two states is perfect (i.e., $(Q^n)_\mathrm{c} = {Q_\mathrm{c}}^n$).

13.5 Error Analysis and a Similarity Metric

The previous sections have explained how to merge pairs of states and have explained that row- or column-equivalent pairs will yield perfect aggregation. Of course, it is highly unlikely that pairs of states will be found that are perfectly row equivalent or column equivalent. The goal then is to find a similarity metric that measures the row and column similarity (i.e., how close pairs of states are to being row or column equivalent). If the metric is formed correctly, those pairs of states that are more similar should yield less error when aggregated. This section will derive an expression for error and then use this as a similarity metric for pairs of states.

We will use $Q_u Q_r$ and ${Q_u}^2$ to estimate error. As mentioned before, it is desirable to have the entries in those two matrices be as similar as possible. Consider aggregating two states i and j. Then the entries in ${Q_u}^2$ are:

$$ {p_{x,y}}^{(2)} = p_{x,i}p_{i,y} + p_{x,j}p_{j,y} + \sum_{k \neq i,j} p_{x,k}p_{k,y} $$

The entries in $Q_u Q_r$ are:

$$ {a_{x,y}}^{(2)} = (p_{x,i} + p_{x,j}) r_{\{i \vee j\},y} + \sum_{k \neq i,j} p_{x,k}p_{k,y} $$

Then the error associated with the (x, y)th element of $Q_u Q_r$ is:

$$ Error_{i,j}(x,y) = {a_{x,y}}^{(2)} - {p_{x,y}}^{(2)} = (p_{x,i} + p_{x,j}) r_{\{i \vee j\},y} - p_{x,i}p_{i,y} - p_{x,j}p_{j,y} $$

Using Eq. 13.1 for $r_{\{i \vee j\},k}$ (and substituting y for k) yields:

$$ Error_{i,j}(x,y) = (p_{x,i} + p_{x,j}) [\frac{m_i p_{i,y} + m_j p_{j,y}}{m_i + m_j}] - p_{x,i}p_{i,y} - p_{x,j}p_{j,y} $$

Now denote $\alpha_{i,j}(y) = p_{i,y} - p_{j,y}$. This is a measure of the row similarity for rows i and j at column y (and will be explained further below). Then:

$$ Error_{i,j}(x,y) = $$
$$ (p_{x,i} + p_{x,j}) [\frac{m_i (p_{j,y} + \alpha_{i,j}(y)) + m_j p_{j,y}}{m_i + m_j}] - p_{x,i}(p_{j,y} + \alpha_{i,j}(y)) - p_{x,j}p_{j,y} $$

This simplifies to:

$$ Error_{i,j}(x,y) = \frac{(m_i p_{x,j} - m_j p_{x,i}) \alpha_{i,j}(y)}{m_i + m_j} $$

Denote $\beta_{i,j}(x) = (m_i p_{x,j} - m_j p_{x,i})/(m_i + m_j)$. Then:

$$ Error_{i,j}(x,y) = \beta_{i,j}(x) \alpha_{i,j}(y) $$

Now $\beta_{i,j}(x)$ can be considered to be a measure of column similarity for columns i and j at row x (this will be shown more explicitly further down). Since only the magnitude of the error is important, and not the sign, the absolute value of the error should be considered:

$$|Error_{i,j}(x,y)| = |\beta_{i,j}(x)\alpha_{i,j}(y)|$$

Recall that $Error_{i,j}(x,y)$ is the error associated with the (x,y)th element of $Q_u Q_r$, if states i and j are aggregated. The total error of the whole matrix is:

$$Error_{i,j} = \sum_x \sum_y |Error_{i,j}(x,y)| = \sum_x \sum_y |\beta_{i,j}(x)\alpha_{i,j}(y)|$$

But this can be simplified to:

$$Error_{i,j} = (\sum_x |\beta_{i,j}(x)|)(\sum_y |\alpha_{i,j}(y)|)$$

To understand this equation consider the situation where states i and j are row equivalent. Then $\forall y \; p_{i,y} = p_{j,y}$. This indicates that $\forall y \; \alpha_{i,j}(y) = 0$ and $Error_{i,j} = 0$. Thus there is no error associated with aggregating row-equivalent states i and j, as has been shown in earlier sections.

Consider the situation where states i and j are column equivalent. Then $\forall x \; p_{x,i} = qp_{x,j}$ and $m_i = qm_j$. It is trivial to show that $\forall x \; \beta_{i,j}(x) = 0$ and as a consequence $Error_{i,j} = 0$. Thus there is no error associated with aggregating column-equivalent states i and j, as has been shown in earlier sections.

Given this, a natural similarity metric is the expression for error:

$$Similarity_{i,j} = (\sum_x |\beta_{i,j}(x)|)(\sum_y |\alpha_{i,j}(y)|) \tag{13.3}$$

If the similarity is close to zero then error is close to zero, and pairs of states can be judged as to the amount of error that will ensue if they are aggregated (it is useful to think of this as a "Dissimilarity" metric). The aggregation algorithm can now be written as follows:

Aggregate()
    Repeat as long as possible
        (i) Find pair of states i and j such that $Similarity_{i,j} < \epsilon$.
        (ii) Aggregate-states(i,j).

The role of ϵ is as a threshold. Pairs of states that are more similar than this threshold can be aggregated. By raising ϵ one can aggregate more states, but with a commensurate increase in error.

This chapter thus far has fully outlined the aggregation algorithm for pairs of states, and identified situations under which aggregation is perfect –

namely, when the pairs of states are row or column equivalent. By performing an error analysis, a natural measure of similarity was derived, in which pairs of states that are row or column similar yield small amounts of error in the aggregation algorithm. The following section outlines some experiments showing the degree of aggregation that can be achieved in practice.

13.6 Some Experiments

In order to evaluate the practicality of the aggregation algorithm, it was tested on some Markov chains derived from the Markov model of an EA presented in Chap. 12. We examine both the error introduced by the aggregation algorithm, and the time taken to aggregate chains of various sizes.

13.6.1 Accuracy Experiments

The first set of experiments examine the accuracy of the aggregated Markov chains by using both ${Q_{\mathrm{u}}}^n$ and ${Q_{\mathrm{c}}}^n$ to compute the probability distribution $p^{(n)}$ over the states at time n. To answer such questions, ${Q_{\mathrm{u}}}^n$ must be combined with a set of initial conditions concerning the EA at generation 0. Thus, the a priori probability of an EA being in state i at time 0 is ${p_i}^{(0)}$ (which is given by Eq. 12.7).[5] Given this, the probability that the EA will be in a particular state j at time n is:

$$ {p_j}^{(n)} = \sum_i {p_i}^{(0)} \; {p_{i,j}}^{(n)} $$

It is also possible to compute probabilities over a set of states. Define a predicate $Pred_J$ and the set J of states that make $Pred_J$ true. Then the probability that the EA will be in one of the states of J at time n is:

$$ {p_J}^{(n)} = \sum_{j \in J} {p_j}^{(n)} $$

As with Chap. 12, J represents the set of all states that contain at least one copy of the optimum (i.e., the set of all populations that have at least one individual with the optimum function value). The Markov model is used to compute ${p_J}^{(n)}$, the probability of having at least one copy of the optimum in the population at time n. The aggregation algorithm can thus be evaluated by using both ${Q_{\mathrm{u}}}^n$ (ground truth) and ${Q_{\mathrm{c}}}^n$ (the estimate) to compute ${p_J}^{(n)}$ for different values of n. The closer the estimate is to ground truth, the better the aggregation algorithm is working.

Since the goal is to compute probabilities involving states containing the optimum (the J set), J states should not be aggregated with non-J states.

[5] If states i and j have been aggregated then ${p_{\{i \vee j\}}}^{(0)} = {p_i}^{(0)} + {p_j}^{(0)}$.

Consequently, the aggregation algorithm is run separately for both sets of states. The algorithm is shown in Fig. 13.1. In theory this aggregation algorithm could result in a two state model involving just J and non-J. In practice this would require large values of ϵ and unacceptable error in ${p_J}^{(n)}$ computations.

Repeat until no new aggregated states are created
  (a) For each state i in the J set of the current aggregated model
    (i) Find the most similar state j in the J set.
    (ii) If $Similarity_{i,j} < \epsilon$, Aggregate-states($i$,$j$).
  (b) For each state i in the non-J set of the current aggregated model
    (i) Find the most similar state j in the non-J set.
    (ii) If $Similarity_{i,j} < \epsilon$, Aggregate-states($i$,$j$).

Fig. 13.1. The final aggregation algorithm

Four different search spaces were chosen for the EA: Type I, NotType I, Type II and NotType II. This particular set of four search spaces was chosen because experience has shown that it is hard to get a single aggregation algorithm to perform well on all. Also, in order to see how well the aggregation algorithm scales to larger Markov chains, four population sizes were chosen for the EA (10, 12, 14, and 16). These four choices of population size produced Markov chains of 286, 455, 680, and 969 states, respectively. Thus, the aggregation algorithm was tested on sixteen different Markov chains.[6]

Naturally, the setting of ϵ is crucial to the success of the experiments. Experiments indicated that a value of 0.15 yielded good aggregation with minimal error, for all sixteen Markov chains. The results for $N = 455$ are shown in Fig. 13.2. The results for the other experiments are omitted for the sake of brevity, but they are almost identical. The values ${p_J}^{(n)}$ are computed for n ranging from 1 to 100, for both the aggregated and unaggregated Markov chains, and graphed as curves.

Table 13.1. The percentage of states removed when $\epsilon = 0.15$

	$N = 286$	$N = 455$	$N = 680$	$N = 969$
Type I	85%	88%	90%	92%
NotType I	65%	73%	79%	82%
Type II	71%	76%	81%	84%
NotType II	64%	73%	79%	82%

The graphs clearly indicate that the aggregated matrix is yielding negligible error. To see how the amount of aggregation is affected by the size of

[6] See De Jong et al. (1994) for a definition of these search spaces.

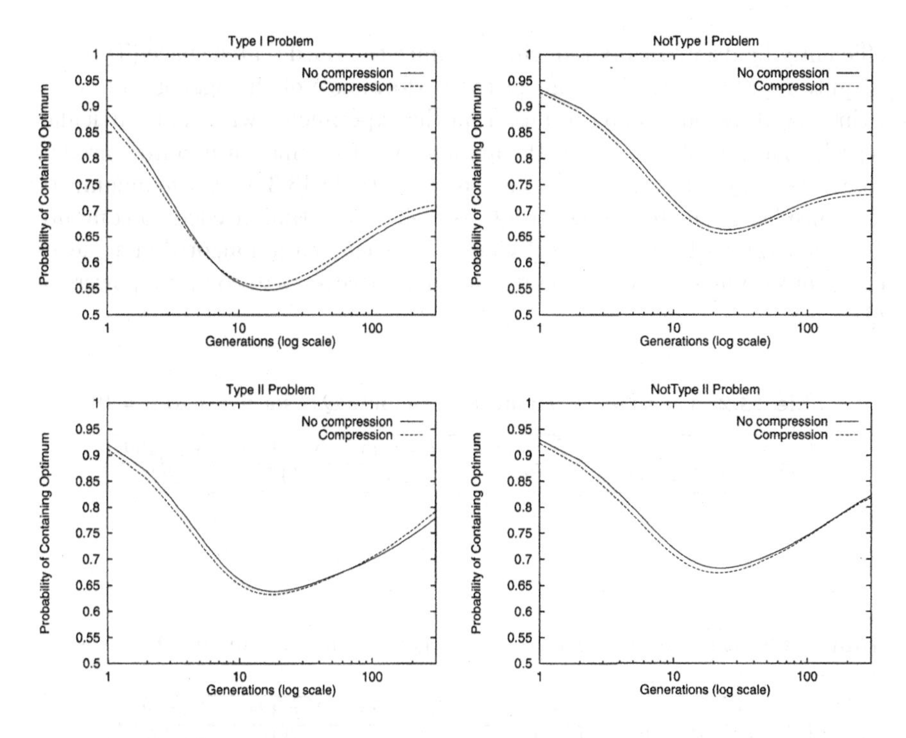

Fig. 13.2. ${p_J}^{(n)}$ where ϵ is 0.0 and 0.15 for $N = 455$. The problems are Type I, NotType I, Type II and NotType II.

the Markov chain, consider Table 13.1, which gives the percentage of states removed for each of the sixteen chains. What is interesting is that, for these particular search spaces, the amount of aggregation is increasing as N increases (while still yielding negligible error). For $N = 969$, over 80% of the states have been removed, yielding Q_c matrices roughly 3% the size (in terms of memory requirements) of the original Q_u matrix.[7] It is also interesting to note that different search spaces are consistently aggregated to different degrees. Further investigation into the nature of these search spaces may help characterize when arbitrary Markov chains are hard/easy to aggregate with this algorithm.

13.6.2 Timing Experiments

It is now necessary to examine the computational cost of the aggregation algorithm. Our prior work (Spears and De Jong 1996) focused heavily on the insights gained by actually examining ${Q_u}^n$, which involved computations on the order of N^3 (to multiply Q_u repeatedly). Thus, the primary motivation for producing the aggregation algorithm was to gain the same insights more

[7] Recent results with $N = 2024$ show even better performance (Spears 1999).

efficiently by dramatically reducing N. Since the third search space (Type II) is quite representative in terms of the performance of the aggregation algorithm, we draw our timing results from the experiments with that particular search space. Table 13.2 gives the amount of CPU time (in minutes) needed to compute ${Q_u}^n$ as n ranges from 1 to 100. Table 13.3 gives the amount of time needed to aggregate Q_u to Q_c as well as the time needed to compute ${Q_c}^n$ as n ranges from 1 to 100.[8] Clearly, the aggregation algorithm achieves enormous savings in time when it is actually necessary to compute powers of Q_u.

Table 13.2. The time (in minutes) to compute ${Q_u}^n$ for $n = 1$ to $n = 100$

	$N = 286$	$N = 455$	$N = 680$	$N = 969$
Computation Time	27	125	447	1289

Table 13.3. The time (in minutes) to aggregate Q_u and to compute ${Q_c}^n$ for $n = 1$ to $n = 100$

	$N = 286$	$N = 455$	$N = 680$	$N = 969$
Aggregation Time	0.2	0.9	3.0	9.5
Computation Time	2.4	7.6	17.9	38.1

Another common use of ${Q_u}^n$ is to compute the probability distribution $p^{(n)}$ over the states at time n (as we did in the previous subsection). If the prior distribution $p^{(0)}$ is known in advance, however, this is more efficiently done by multiplying $p^{(0)}$ by Q_u repeatedly (i.e., this is repeated n times to produce $p^{(n)}$). The computation is of order N^2 instead of N^3.

Table 13.4. The time (in minutes) to compute $p^{(n)}$ for $n = 1$ to $n = 100$

	$N = 286$	$N = 455$	$N = 680$	$N = 969$
Computation Time	0.1	0.3	0.7	1.4

Table 13.4 and Table 13.5 give the amount of time needed to compute $p^{(n)}$ (from Q_u and Q_c respectively). Despite the obvious benefits of computing $p^{(n)}$ from Q_c, the aggregation algorithm is not advantageous in this case since the time needed to aggregate Q_u exceeds the time to produce $p^{(n)}$ from Q_u. However, there are still occasions when aggregating Q_u and then using Q_c to

[8] All timing results are on a Sun Sparc 20. The code is written in C and is available from the author.

Table 13.5. The time (in minutes) to aggregate Q_u and to compute $p^{(n)}$ for $n = 1$ to $n = 100$

	$N = 286$	$N = 455$	$N = 680$	$N = 969$
Aggregation Time	0.20	0.90	3.00	9.50
Computation Time	0.02	0.02	0.03	0.05

compute $p^{(n)}$ will in fact be more efficient. The first is when it is necessary to compute $p^{(n)}$ for a large number of different prior distributions (recall that Q_c does not depend on the prior information and hence need not be recomputed). The second occasion is when it is necessary to compute $p^{(n)}$ for large n (e.g., Stewart 1994 indicates that times on the order of 10^8 are sometimes required). In both of these situations the cost of the aggregation algorithm is amortized. Finally, aggregation is also advantageous when the prior distribution is not known in advance.[9]

In summary, the aggregation algorithm is most advantageous when it is necessary to actually examine the powers of Q_u directly. For computing probability distributions over the states, the aggregation algorithm will be advantageous if the prior distribution is initially unknown, if a large number of prior distributions will be considered, or if the transient behavior over a long period of time is required.

13.7 Related Work

The goal of this chapter has been to provide a technique for aggregating (or *compressing*) discrete-time Markov chains (DTMCs) in a way that yields good estimates of the transient behavior of the Markov model. This section summarizes the work that is most closely related.

There is a considerable body of literature concerning the approximation of transient behavior in Markov chains. Techniques include the computation of matrix exponentials, the use of ordinary differential equations, and Krylov subspace methods (Stewart 1994). However, all of these techniques are for continuous-time Markov chains (CTMCs), which use an infinitesimal generator matrix instead of a probability transition matrix. It is possible to discretize a CTMC to obtain a DTMC such that the stationary probability vector of the CTMC is identical to that of the DTMC. However, Stewart (1994) notes that the transient solutions of DTMCs are not the same as those of the corresponding CTMCs, indicating that these techniques will be problematic for computing the transient behavior of DTMCs.

There is also considerable work in aggregation of DTMCs. Almost all theoretical analyses of aggregation (e.g., the "block aggregation" of Kemeny

[9] It is also important to emphasize that it is very likely that the aggregation algorithm can be extensively optimized, producing much better timing results.

and Snell 1960) utilize the same functional form:

$$f(Q_{\mathrm{u}}) = Q_{\mathrm{c}} = AQ_{\mathrm{u}}B \quad \text{such that } AB = I$$

where A and B are matrices that determine the partitioning and the aggregation of the states (Howe and Johnson 1989a; Howe and Johnson 1989b). This functional form must satisfy two axioms: "linearity" and "state partitioning." Linearity implies that A and B do not depend explicitly on the entries in Q_{u}. State partitioning implies that the "aggregated" transition probabilities should depend only upon the probabilities associated with the aggregated states (e.g., the aggregation of states i and j should only depend on $p_{i,i}$, $p_{i,j}$, $p_{j,i}$, and $p_{j,j}$).

Neither axiom is true for aggregation of column-equivalent states in this chapter. This is reflected in the fact that in general $AB = XY \neq I$. Instead, in this chapter $BA = I$ for both row and column equivalence, yielding desirable properties with respect to the powers of Q_{u}. The current results indicate that the relevance of both axioms should be re-examined.

The aggregation technique most closely related to the work in this chapter is described by Stewart (1994), Stewart and Wu (1992), and Vose (1995). This aggregation technique partitions the set of states S into s nonempty sets $S_1, \ldots, S_s$. Denoting the steady-state probability of state i as π_i, then $\pi_y = \sum_{i \in S_y} \pi_i$ if:

$$Q_{\mathrm{c}}(x,y) = \frac{1}{\sum_{i \in S_x} \pi_i} \sum_{i \in S_x} \left[\pi_i \sum_{j \in S_y} p_{i,j} \right] \tag{13.4}$$

If aggregation is performed in this manner, the steady-state behavior of the aggregated system is the same as the original system. The aggregated matrix can be computed via the method of "stochastic complementation" or via "iterative aggregation/disaggregation" methods. The former will work on arbitrary matrices but is generally computationally expensive. The latter is most efficient for "nearly completely decomposable" (NCD) matrices (Dayar and Stewart 1997). However, the emphasis is always on steady-state behavior, and not on transient behavior. This difference in emphasis can been seen by noting the difference in the choice of weights (compare Eq. 13.2 to Eq. 13.4) – the focus in this chapter has been on column mass instead of steady-state values.

In a sense the aggregation algorithm presented in this chapter is a generalization of steady-state aggregation. The steady-state matrix is column equivalent for every pair of states, and the column masses, when renormalized, are the same as the steady-state probabilities. Thus the aggregation algorithm is a generalization of the steady-state aggregation formula to transient behavior.[10] This leads to the intriguing hypothesis that this new aggregation

[10] Note that Lemma 13.4.3 implies that if $\mathbf{b}_i = q\mathbf{b}_j$ for states i and j in Q_{u}, then $\pi_i = q\pi_j$.

algorithm will be more accurate when describing transient behavior, and less accurate for describing steady-state behavior. Preliminary results appear to confirm this hypothesis.

13.8 Summary

This chapter has introduced a novel aggregation algorithm for probability transition matrices. The output from the algorithm is a smaller probability transition matrix with fewer states. The algorithm is designed to aggregate arbitrary (not necessarily NCD) probability transition matrices of DTMCs in order to obtain accurate estimations of transient behavior. Thus it appears to fill the gap between existing transient techniques (which focus on CTMCs) and existing aggregation techniques for DTMCs (which focus on steady-state behavior).

There are a number of potential avenues for further expansion of this work. The first possibility is to aggregate more than two states at once. Multiple-state aggregation may yield better results, by allowing for a more accurate estimation of error. Another avenue is to derive estimates of how error propagates to higher powers of Q_c. The current similarity metric is not necessarily a good indicator of the error at higher powers of Q_c, although empirically the results are quite good. However, both of these avenues greatly increase the computational complexity of the algorithm.

The comparison with the related work indicates that this new aggregation algorithm can be considered to be a generalization of the more traditional aggregation formulas. This indicates yet a third avenue for research. If in fact column mass turns out to yield better weights for the weighted average during transient behavior, then it may be possible to smoothly interpolate between column mass and steady-state probabilities as the transient behavior approaches steady state. Of course, this presupposes the existence of the steady-state distribution, but efficient algorithms do exist to compute these distributions.

The current algorithm also quite deliberately ignores the roles of the priors ${p_i}^{(0)}$, in order to have as general an algorithm as possible. However, if priors are known, then it may be possible to use this information to improve the weighted averaging procedure (see Appendix), thus once again reducing the error in some situations.

Finally, the amount of aggregation that can be achieved with negligible error is a useful indicator of whether the system is being modeled at the correct level of granularity. If the probability transition matrix is hard to aggregate, then the system is probably modeled at a reasonable level of granularity. However, ease of aggregation indicates that the system is being modeled in too much detail. In these cases monitoring the states that are chosen for aggregation by the similarity metric can yield important information about the characteristics of the system. This approach could be used to characterize systems that are defined by a probability transition matrix but are still not well understood at a higher level.

This completes the excursion into aggregating Markov models. Our main concern now is in what Markov models can teach us about EAs, especially with respect to the roles of recombination and mutation. As mentioned in the prior chapter, which used Markov chain models of EAs (Chap. 12), our initial progress suggests that the multimodality of (the number of peaks in) the search space is of importance in determining the relative performance of recombination and mutation. This observation will now be confirmed by using real EAs, in Chap. 14.

Part IV

Empirical Analyses

14. Empirical Validation

14.1 Introduction

The results of the earlier static schema analyses from Chap. 8 and of the dynamic Markov chain analyses from Chap. 12 strongly suggested that the multimodality of (number of peaks in) a search space is an important characteristic in determining the relative importance of mutation and recombination in an EA. The results from the Markov chain analyses also suggested that the relative heights of the peaks had a strong effect on the performance of EAs with recombination. Naturally, it is important to empirically validate these results with *real* EAs, in order to see if the results scale to larger, more realistic problems. How should this empirical validation be performed?

One weakness of standard empirical studies in which search algorithms are compared is that their results may not generalize beyond the test problems used. A classic example of this is a study in which a new algorithm is carefully tuned to the point that it outperforms some existing algorithms on a few ad hoc problems (e.g., the De Jong 1975 test suite). The results of such studies typically have only weak predictive value regarding relative performance on new problems.

There are two ways to strengthen the results obtained from empirical studies. As was discussed in Chap. 2, the first is to remove the opportunity to hand-tune algorithms to a particular problem or set of ad hoc problems. This can be done by using "test-problem generators," which produce random problems from within a well-specified class of problems. Having problem generators allows one to report results over a randomly generated set of problems that have well-controlled characteristics, rather than a few hand-chosen, ad hoc examples. Thus, by increasing the number of randomly generated problems, the predictive power of the results for the problem class as a whole has increased. An advantage of problem generators is that in most cases they are quite easy to parameterize, allowing one to design controlled experiments in which one or more properties of a class of problems can be varied systematically to study the effects on particular search algorithms.

On a related issue, it is common practice to run EAs to some fixed termination criteria, and then to report the results only after termination. However, this ignores the dynamic aspects of an EA, and can lead to overly general conclusions. For example, as we will see, conclusions can often turn out to be surprisingly dependent on the termination criteria, often reversing if a different cutoff is used. From both an engineering and scientific standpoint it is crucial to include results throughout the running of EAs. Thus, a second way to improve empirical methodology is to always show results over the whole running time of an EA (as was done in Chap. 12).

This chapter compares the performance of recombination and mutation in *real* EAs using a problem generator motivated by the results from Chap. 12 of this book. The goal is to explore the behavior of recombination and mutation as various aspects of the search space are methodically changed.

14.2 The Multimodal Problem Generator

The results from Chap. 12, which used a Markov chain analysis of an EA, strongly suggested that the multimodality of (number of peaks in) a search space is an important characteristic of any search space. However, since those results were obtained from very small problems (with two- and three-bit individuals and small population sizes), there is naturally some concern as to how these results scale to larger problems and larger population sizes. To this end we create a test-problem generator that generates large random problems with a controllable degree of multimodality.

The Markov chain analysis of Chap. 12 was performed on small one-peak and two-peak problems. To understand the motivation for examining multimodality, consider a simple two-peak problem, with optima at 000...000 and 111...111. Individuals (strings) with roughly 50% 1s and 0s are the lowest fitness strings, while individuals with mostly 1s or mostly 0s have high fitness. Mutation of any high-fitness individual on either peak will tend to keep the individual on that peak, driving it up or down the peak to a small degree. Recombination, however, produces quite different results, depending on the location of the parents. If the two parents are on the same peak, the offspring are also highly likely to be on that peak. However, if the two parents are on the two different peaks, the offspring are highly likely to be in the valley between the two peaks, where the fitness is low. The results in Chap. 12 confirm this hypothesis for very small problems, by showing that an EA with recombination outperforms an EA without recombination on small one-peak problems, whereas an EA without recombination outperforms an EA with recombination on small two-peak problems.

What if there are more than two peaks? It appears reasonable to hypothesize that recombination could be even more deleterious, since the recombination of individuals on different peaks is even more likely to produce poor offspring, until the population has converged to one peak. This hypothesis is consistent with the view of Jones (1995) that fitness functions should be considered as "operator landscapes," i.e., they should be considered from an operator point of view. Our notion of multimodality is generally embedded in Hamming space, which is an ideal view for mutation, since the mutation of a parent yields a child that is nearby in Hamming space. Thus we might expect mutation to perform well on multimodal functions. However, Hamming distance is not necessarily useful when considering recombination, since the recombination of two parents can yield children arbitrarily far in Hamming space (i.e., in the valleys between two peaks). Thus again we might expect recombination to perform poorly on multimodal functions.

To explore these hypotheses, a multimodality problem generator was created, in which the number of peaks $\mathcal{P}$ (the degree of multimodality) can be controlled easily and methodically by the experimenter. The idea is to generate a set of $\mathcal{P}$ random L-bit strings, which represent the location of the $\mathcal{P}$ peaks in the space. To evaluate an arbitrary binary string, first locate the nearest peak (in Hamming space). Then the fitness of the binary string is the number of bits the string has in common with that nearest peak, divided by L.

$$f(string) \;=\; \frac{1}{L} \max_{i=1}^{\mathcal{P}} \{L - \text{Hamming}(string, Peak_i)\}$$

14.2.1 Experiments with All Peaks at Equal Heights

The results from Chap. 12 suggested that recombination may perform worse as the number of peaks in a space increases. However, these results were generated from very small problems. The multimodality generator defined above provides a nice mechanism for investigating this result further, by allowing the user to create large random problems with a desired degree of multimodality. What one would expect to see is a gradual degradation in performance of an EA with recombination, as the number of peaks in the space increases.

Of the EAs, both "evolution strategies" (Rechenberg 1973; Schwefel 1981), and "evolutionary programming" (Fogel et al. 1966) are most often used for real-valued problems, whereas "genetic algorithms" (Holland 1975) are most often used for discrete problems. Since our problem generator is defined over binary strings, the natural choice for our EA is a genetic algorithm (GA). The GA chosen (called "GAC," which is available from the author), is quite traditional, with fitness-proportional selection, mutation and recombination. Furthermore, GAC is also quite similar to the simple EA assumed in Chap. 12 – its only differences are in its use of fitness "scaling" (see Goldberg 1987) and the fact that recombination produces two offspring as opposed to one. Neither of these differences should produce qualitative differences in results.

To see how multimodality affects recombination and mutation we ran the GA in three different modes. In the first mode both mutation and recombination are used. In the second only mutation is used. In the third only recombination is used. Fitness-proportional selection is always used. The expectation is that the two GAs with recombination should perform worse as the number of peaks increases. To test this we created problems ranging from 1

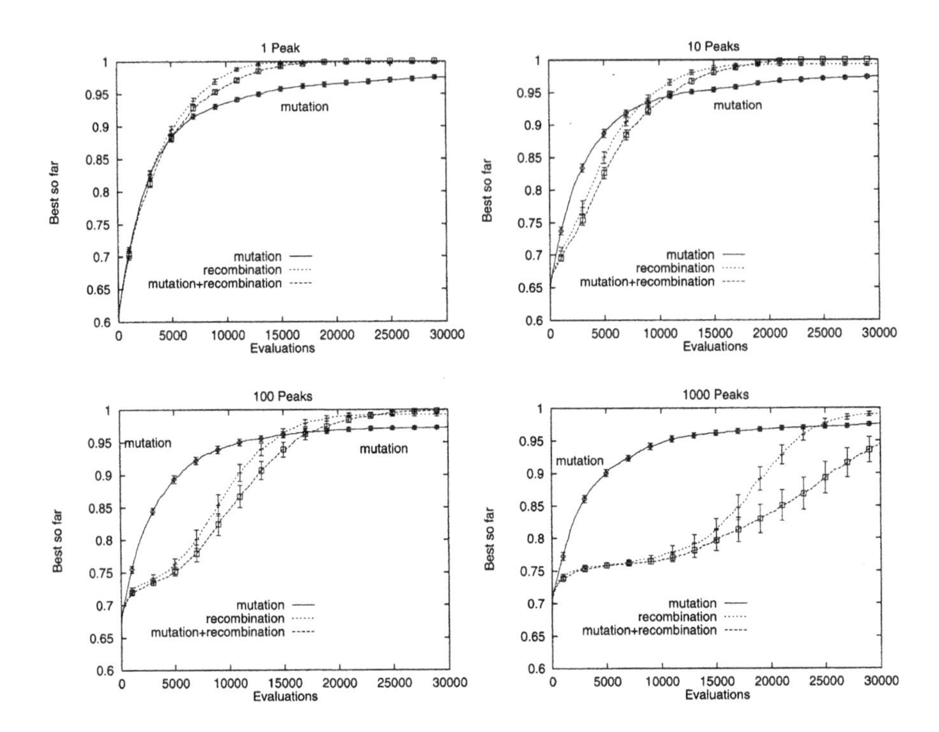

Fig. 14.1. Best-so-far performance of an EA on 1-peak, 10-peak, 100-peak, and 1000-peak problems

peak to 1000 peaks. For a given number of peaks, fifty random problems were created. The GA was run once per problem, and the results were averaged over those fifty problems. The string length L was 100, the population size P was 100, the mutation rate $\mu = 0.001$ and the recombination rate $\chi = 0.6$ (this is the percentage of individuals that are recombined every generation, when recombination is turned on). These are quite typical parameter settings for GAs – e.g., see De Jong (1975). The GA was run for 30,000 evaluations, where an individual was evaluated only if it differed from its parent.

The performance metric that was monitored is also quite traditional – namely "best-so-far" curves that plot the fitness of the best individual that has been seen thus far by generation n. Since the multimodality generator produces problems that range from 0.0 to 1.0 in fitness, the best-so-far curves will fall within those fitness values. Since higher fitness indicates being closer to an optimum, our GA will have to maximize the function.

Figure 14.1 illustrates the results, which are quite striking. The behavior of the GA without recombination (i.e., with selection and mutation) is almost independent of the number of peaks. This seems reasonable, since the presence of multiple peaks will not influence the mutation of an individual on a particular peak. Recombination, on the other hand, performs quite differently.[1] The GAs with recombination outperform the GA without recombination on the one-peak problems. However, what is most noticeable is the severe drop in performance of the GAs with recombination as the number of peaks increases. This is consistent with our hypothesis – simply put, recombination of individuals on multiple peaks can often produce poor performing individuals in the valleys between peaks. Note, however, that eventually the performance curves for the GAs with recombination pick up dramatically in later generations. What appears to be happening is that by this point the population has lost so much diversity that in fact the individuals are clustered around one peak. At this point recombination becomes beneficial and performance increases.

14.2.2 Experiments with Peaks at Unequal Heights

The previous subsection shows results that confirm the hypothesis (from Chap. 12) that increasing multimodality can have a deleterious effect on EAs with recombination. As expected, as the number of peaks increased, the performance of recombination worsened.

In the previous subsection, all of the $\mathcal{P}$ peaks had the same maximum height of 1.0. However, Chap. 12 also indicated that lowering the height of a suboptimal peak tended to help the performance of recombination (i.e., this tends to make the problem look more like a one-peak problem). This effect can be tested with the multimodality generator by explicitly adding a new control knob to the generator that can change the height of peaks. This is done by assuming that the first peak has height 1.0 while peak $\mathcal{P}$ has some height h lower than 1.0. The remainder of the peaks have heights linearly interpolated between h and 1.0. Thus, the user can modify the maximum height on each peak by changing h. For example, in a problem with four peaks, having $h = 0.0$ means that the first peak has height 1.0, the second peak has height 2/3, the third peak has height 1/3, and the fourth has height 0.0.[2]

To test the effect that the height of suboptimal peaks has on recombination, we produced fifty random 10-peak problems, for each of the four different settings of h: 1.00, 0.66, 0.33, 0.00. The parameter settings of the GA are the same as in the last subsection. The expectation is that although recombination has difficulty with a larger number of peaks, the difficulty should get less as h is reduced. Figure 14.2 shows the best-so-far results, which confirm our expectations. Note that a small reduction in the height of the peaks does help recombination noticeably.

[1] We used one-point recombination throughout this chapter, but the results are similar for two-point and 0.5 uniform recombination. The vertical bars overlaying the best-so-far curves represent 95-percent confidence intervals computed from Student's t-statistic (Miller 1986).

[2] The fitness of an individual near peak i is scaled by the height of peak i.

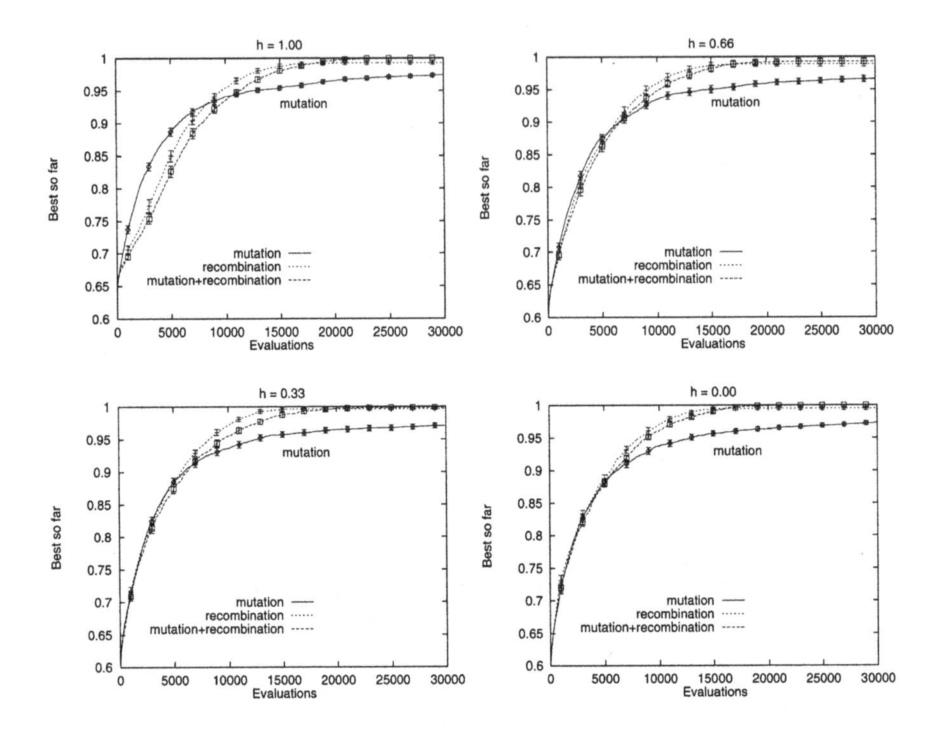

Fig. 14.2. Performance of an EA on 10-peak problems where the minimum height h of a peak ranges from 1.0 to 0.0

To see how the multimodality affects these results, we also performed the same experiment with 100-peak problems. Figure 14.3 shows the results. The effect is still quite dramatic in that lowering the height of the peaks does help recombination noticeably.

14.3 The Relationship of Multimodality to Epistasis

By use of a multimodality problem generator, the previous section has confirmed that the effectiveness of recombination is intimately connected to the number of peaks in a space, as well as heights of those peaks. However, according to Chaps. 1 and 2, another characteristic of a space that may prove important for recombination is called "epistasis." Is it possible to find a relationship between these two different concepts?

As pointed out in Chap. 8, recombination has a clear advantage (over mutation) in terms of its ability to construct higher-order building blocks from lower-order building blocks. This has led Fogel (1995) to hypothesize that recombination will perform poorly for most naturally evolved systems, because (so he claims) they are extensively pleiotropic (a gene may influence

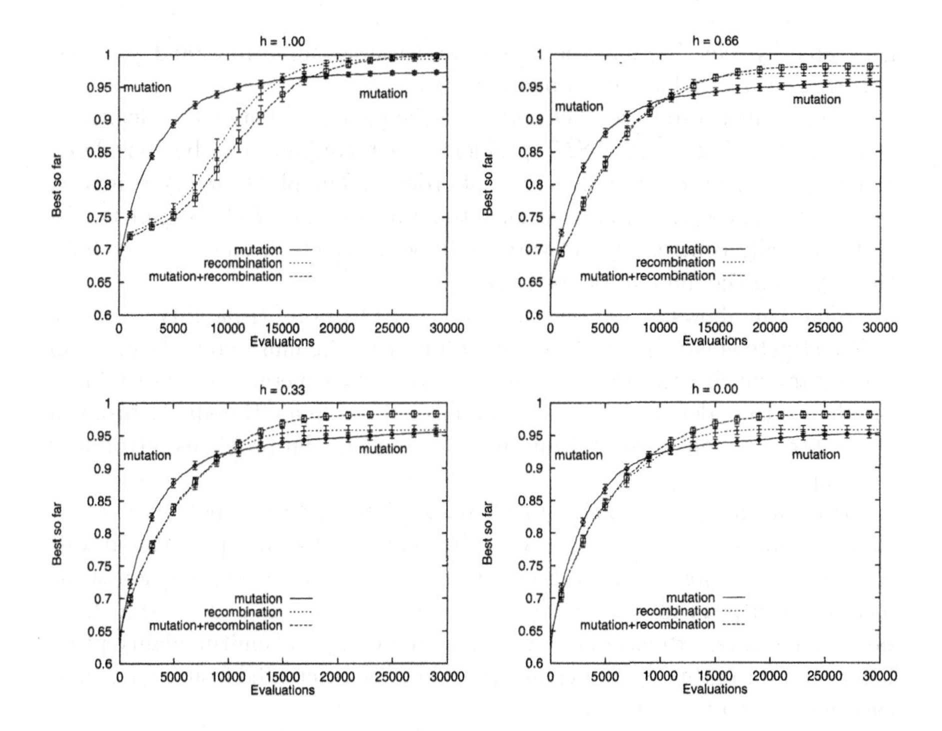

Fig. 14.3. Performance of an EA on 100-peak problems where the minimum height h of a peak ranges from 1.0 to 0.0

multiple traits) and highly polygenic (a trait may be influenced by multiple genes). Since such systems will not have many high-fitness building blocks for recombination to exploit, Fogel argues that mutation will be superior for these systems.

The biological concepts of pleiotropy and polygeny are related to "epistasis." A system has low (high) epistasis if the optimal allele for any locus depends on a small (large) number of alleles at other loci. Systems with independent loci (the optimal allele for each locus can be decided independently of the alleles at the other loci) have no epistasis.

Chapter 2 investigated the application of EAs to solving Boolean satisfiability problems and provided a mapping from arbitrary Boolean satisfiability problems to mathematical fitness functions amenable to solution by EAs. As shown in that chapter, it is easy to construct satisfiability problems with a controlled amount of epistasis. This technique was pursued in more detail in De Jong et al. (1997), which examined a test-problem generator referred to as "Random $\mathcal{L}$-SAT" (Mitchell et al. 1992). The Random $\mathcal{L}$-SAT problem generator creates random Boolean expressions in conjunctive normal form (CNF) subject to the three parameters $\mathcal{V}$, $\mathcal{C}$, and $\mathcal{L}$. Each of $\mathcal{C}$ conjuncts is generated by selecting $\mathcal{L}$ of the $\mathcal{V}$ Boolean variables uniformly randomly and negating each variable with probability 0.5.

We can make direct contact here with the biological notions of pleiotropy and polygeny. For these $\mathcal{L}$-SAT problems, each conjunct can be considered to be a trait. Hence, the polygeny is of order $\mathcal{L}$. The pleiotropy is estimated by noting that each variable occurs (on average) in $\mathcal{CL}/\mathcal{V}$ conjuncts. By systematically controlling and varying these parameters, one can vary both the type and the amount of epistasis.

What this shows is that one can use a generator of satisfiability problems to investigate epistasis. This provides a link with the multimodality problem generator, which can also be considered to be a generator of satisfiability problems. To understand this one has to see that the multimodality problem generator is simply a generalization of the multimodal problems introduced in Chap. 2.

The multimodal problems in Chap. 2 (where the number of peaks $\mathcal{P}$ ranged from one to six) are based on Boolean satisfiability problems in disjunctive normal form (DNF). Each disjunct of the Boolean expression is the location of a peak in the search space (i.e., there are as many disjuncts as peaks). Thus, by extension, any problem created by the multimodality problem generator in this chapter can be considered to be created from a particular satisfiability problem in DNF.

Thus what we have shown is that the Random $\mathcal{L}$-SAT generator, which creates satisfiability problems in CNF, can be used to investigate epistasis. The multimodal problem generator, which is used to investigate multimodality, creates satisfiability problems in DNF. Any problem in CNF can be converted to DNF and vice versa – thus the framework of satisfiability allows us to link the notions of epistasis and multimodality.

In general, increasing the number of disjuncts in a satisfiability problem (in DNF) will increase the number of conjuncts (and the complexity of the conjuncts) in the equivalent expression in CNF. Increasing the number of peaks in the multimodal problems will tend to increase their epistasis. Our results show a degradation in the performance of recombination as the multimodality increases, which is consistent with Fogel's hypothesis.

However, consider the experiments in which the fitness of the peaks is methodically lowered. Changing the fitness does not change the syntactic expression of the multimodal function as a DNF Boolean expression. Thus changing the fitness of the peaks will also not change the epistasis (as measured syntactically in terms of the number of conjuncts and the complexity of the conjuncts) of the equivalent CNF Boolean expression. Yet, despite the lack of change in epistasis, the performance of recombination is dramatically affected. This illustrates the danger in attempting to characterize fitness functions with purely syntactic measures that ignore fitness to a large extent. Syntactic measures of epistasis are simply inadequate. In order for such characterizations of fitness functions to be useful, they must include fitness information to a much larger extent (e.g., see Davidor 1990).

14.4 Summary

This chapter emphasizes the empirical methodology of using "test-problem generators" to produce random problems from within a well-specified class of problems. Having problem generators allows one to report results over a randomly generated set of problems that have well-controlled characteristics, rather than a few hand-chosen, ad hoc examples. The goal is to identify problem characteristics that yield useful predictive theories concerning performance.

The results of the earlier static schema analyses from Chap. 8 and of the dynamic Markov chain analyses from Chap. 12, strongly suggested that the multimodality of (number of peaks in) a search space, as well as the relative heights of the peaks, are important characteristics in determining the relative importance of mutation and recombination in an EA. In order to address these issues, a multimodality test-problem generator was created that allows the user to methodically control both the number of peaks and the relative heights of those peaks in a search space. Results confirm that when all peaks have equal heights, increasing the number of peaks has increasing deleterious effects on the performance of EAs with recombination. However, gradually lowering the heights of the suboptimal peaks is beneficial to the performance of recombination. Interestingly, the EA with mutation (and no recombination) was almost completely unaffected by these changes.[3]

This chapter further shows that the concept of "epistasis" can be linked to the concept of multimodality by using the common language of Boolean satisfiability problems. Multimodal functions can be easily represented as Boolean expressions in disjunctive normal form, while epistasis is easily represented as Boolean expressions in conjunctive normal form. The results presented in this chapter indicate that such syntactic measures of epistasis are inadequate in that they do not include fitness information sufficiently.

[3] Further experiments that compare EAs with the "particle swarm" algorithm on problems drawn from the multimodality test-problem generator can be found in Kennedy and Spears (1998).

Part V

Summary

15. Summary and Discussion

15.1 Summary and Contributions

The central theme of this book was a theoretical and empirical study of recombination and mutation in EAs with the objective of better characterizing the roles of these operators. This theme proceeded in stages. First, static, component-wise analyses of recombination and mutation were performed in isolation. Then, dynamic analyses were performed, which included all aspects of an EA. The results from the static analyses were used to drive the experiments performed in the dynamic analyses. Finally, the results from both the static and dynamic analyses were confirmed empirically with real EAs. This occupied Chaps. 3–8, as well as Chaps. 12 and 14.

The book also made occasional excursions. The purpose of the excursions was to introduce new techniques for studying EAs, as well as to unify the current techniques more tightly by showing the explicit connections between them.

15.1.1 The Central Theme

The central theme began with a static analysis of the effect that recombination and mutation have on kth-order hyperplanes H_k. The analysis was static because it did not take into account the time evolution of the population. In order to provide a fair comparison of recombination and mutation, both operators were treated as two-parent operators. The "disruptive" and "constructive" aspects of both operators were compared by calculating the expected number of offspring that will be in H_k after the two parents have been recombined or mutated. This framework allows for a natural treatment of arbitrary cardinality alphabets, arbitrary population diversity (homogeneity), and arbitrary order hyperplanes.

The results from that static analysis indicated that mutation is more powerful than recombination in terms of disruption – mutation can achieve the same low levels of disruption that recombination can, but can also achieve higher levels of disruption. On the other hand, recombination is more powerful than mutation in terms of construction – recombination has a higher likelihood of constructing two lower-order, nonoverlapping hyperplanes into a higher-order hyperplane than does mutation. The constructive advantage of recombination is maximized when the two lower-order hyperplanes are roughly half the order of the higher-order hyperplane being constructed. The constructive advantage translates into a performance advantage when constructed higher-order hyperplanes have high fitness. On the other hand, recombination should have a deleterious effect on an EA when the higher-order hyperplanes that are constructed have low observed fitness. It was then hypothesized that multimodal (multiple-peak) fitness functions should show that deleterious effect – recombination should perform worse as the number of peaks increases since the recombination of individuals on different peaks will be likely to produce offspring in the valleys between the peaks.

In order to test this hypothesis, dynamic analyses of EAs were investigated in which the time evolution of the EA was explicitly considered. The Nix and Vose (1992) Markov chain model of a complete, simple EA was used to analyze the behavior of an EA on very small problems with one and two peaks. The results confirmed that the multimodality of (number of peaks in) a search space is an important factor in determining the utility of recombination in an EA – recombination was more useful than mutation on the one-peak problems, but less useful than mutation on the two-peak problems. The results also suggested that the relative heights of the peaks influenced the utility of recombination.

In order to confirm these results with real EAs on real problems, the book then introduced an empirical methodology based on "test-problem generators." Test-problem generators can create random problems from a certain class of problems with user-controlled characteristics. A multimodality test-problem generator was created that allows the user to methodically control both the number of peaks and the relative heights of those peaks in a search space. An actual EA was run on large problems drawn from the multimodality generator. Results confirmed that when all peaks have equal heights, increasing the number of peaks has an increasingly deleterious effect on the performance of EAs with recombination. However, gradually lowering the heights of the suboptimal peaks is beneficial to the performance of recombination. Interestingly, the EA with mutation (and no recombination) was almost completely unaffected by the number of peaks or their heights. These results concluded the central theme.

15.1.2 Excursions

As well as following the central theme, the book also took occasional excursions into related theoretical areas. The first excursion was a formal investigation of an observation made in the static schema analysis of the central theme – that more disruptive recombination operators also tend to be more constructive. The book showed that this relationship is always true by proving a "No Free Lunch" theorem which states that any increase in disruptive ability by a recombination operator is matched by an equal increase in constructive ability. In general, there is no such theorem for mutation.

The book then investigated other static characterizations of recombination and mutation, namely, their "exploratory power," their "positional bias," and their "distributional bias." In general, mutation has greater exploratory power than recombination. Like P_0 uniform recombination, mutation has no positional bias, and its distributional bias is also most similar to that of uniform recombination. Interestingly, it was shown that when the cardinality C of the alphabet is two and the population is maximally diverse, mutation and P_0 uniform recombination have the same exploratory power, positional bias, and distributional bias. Surprisingly, this relationship also holds for the static *schema* analysis – when $C = 2$ and the population is maximally diverse, mutation and uniform recombination have precisely the same disruptive and constructive effects on all hyperplanes if $\mu = P_0$.

After the static analyses of recombination and mutation, the book explored dynamic analyses. The first such dynamic analysis concerned the evolution of a population undergoing recombination and/or mutation (but without selection). Previous results indicated that a population undergoing only recombination approaches Robbins' equilibrium. This book showed that a population subject to mutation approaches a "uniform" equilibrium. This equilibrium takes precedence when *both* mutation and recombination act on a population because mutation actually moves Robbins' equilibrium to the uniform equilibrium. The book also attempted to characterize the speed at which these limiting distributions are approached. For recombination it was possible to demonstrate that there are strong connections to the earlier static schema analysis, because in many situations the more disruptive recombination operators drive a population to Robbins' equilibrium more quickly. Similarly the more disruptive mutation is, the more quickly it drives a population towards the uniform equilibrium.

The book then developed a dynamic model of an EA that includes selection and mutation. In general, this involves the iteration of a large number of equations of motion. However, the book then defined a class of fitness functions under which a useful aggregation of the model can be applied, resulting in far fewer equations. This class of functions includes unimodal functions from the EA and biology communities, two-peak deceptive problems from the GA community, and multimodal functions. Since some EAs do not use recombination, this particular model could be a quite valuable theoretical tool.

Unfortunately, as recombination is added to the previous model, the model becomes far more complex. Aggregation also becomes far more difficult to perform. Thus the book explored the possibility that complex models can be automatically aggregated into simple models. The result was a novel aggregation algorithm that can be applied to Markov chain models of complex systems (such as the Nix and Vose model of an EA). The algorithm aggregates a Markov chain into a much smaller Markov chain that is easier to analyze. Preliminary tests of this aggregation algorithm indicated that substantial amounts of aggregation can be performed while introducing only small amounts of numerical error. This particular excursion has scope well beyond that examined in this book, since Markov chains are a common technique for modeling complex systems (and not just EAs).

15.2 Future Work

There are a large number of possible extensions to the work presented in this book. Since the book focused on fixed-length, linear chromosomes, one of the most obvious extensions is to variable-length, nonlinear representations. One example of work in this area is by O'Reilly (1994), which analyzes the Lisp representation of "genetic programming" (Koza 1992).

In general, the static analyses of mutation and recombination in this book examined each operator in isolation. However, it should be possible to extend these analyses to include the behavior of mutation and recombination in combination. For example, it should not be too difficult to calculate the expected number of offspring that will be in a hyperplane H_k after the two parents have been recombined *and* mutated. Likewise it should not be difficult to determine the exploratory power, distributional bias, and positional bias of mutation and recombination both acting on the same parents.

As for the dynamic analyses, much more work remains to be done with respect to the speed at which different recombination operators approach Robbins' equilibrium. The analysis presented in this book allows one to make relative statements concerning the different speeds of different recombination operators, but clearly it would be advantageous to be able to make absolute statements. For example, and stated somewhat simplistically, it would be desirable to be able to compute the "half-life" of a recombination operator, i.e., the time it would take a recombination operator to drive a population half-way to Robbins' equilibrium.

This book has presented one novel test-problem generator for multimodal functions. This particular test-problem generator was very useful in creating a problem class that an EA with recombination has great difficulty solving. Clearly this is not the only possible problem generator, and much work needs to be done in finding other problem characteristics that influence EAs strongly. One potentially useful problem generator would be one that can generate problems that are easy for an EA with recombination but very hard for an EA with only selection and mutation. The class of Hamiltonian circuit problems introduced in Chap. 2 provide a lead in this direction, since recombination was very important for successful search on these problems.

Since it may be difficult to identify important problem characteristics a priori, it is possible that an automated technique can help in this endeavor. Imagine having a meta-search algorithm that searches the space of problems (functions), in an attempt to find problems (functions) that are hard or easy for a particular algorithm (optimizer). Such a technique was used in De Jong et al. (1994) to find a hard problem for an EA. Then, given examples of hard and easy problems, the important problem characteristics may be much easier to discern.

Finally, once important problem characteristics are known, it will be crucial to determine whether problems (functions) have those characteristics. One possible mechanism for achieving this is through "probes" that quickly sample a function to estimate its characteristics, allowing the user to match the problem to the right algorithm.

For example, for a class of Boolean satisfiability problems in conjunctive normal form (Mitchell et al. 1992), the ratio of the number of conjuncts to the number of Boolean variables in the satisfiability problem is an important problem characteristic that often helps determine the best algorithm to apply to that problem. The probe consists of a linear pass over the Boolean expression, computing the conjunct to variable ratio. Similarly, future work in EAs should concentrate on creating probes that quickly measure useful problem characteristics for various problem classes, allowing the user to match algorithms (or operator settings within an algorithm) to particular problems.

15.3 Conclusions

As mentioned in Chap. 1 of this book, one criticism that is often levied against the static "schema" theories is that they are too simple to be useful, since such component-wise analyses cannot be sufficiently predictive. At the opposite extreme, Markov chain models of EAs (although predictive in nature) are considered to be problematic due to their computational complexity. Since they can only be applied to very small problems, there is the added concern that any results obtained with the small problems will not scale to larger, more realistic problems.

This book addressed both criticisms. Markov theories can in fact provide quite useful insights – the behavior of an EA on small (computationally tractable) problems can in fact be observed in larger problems (Chap. 12 and Chap. 14). Furthermore, as mentioned above, it is possible to provide automatic tools for simplifying these models to make them far more computationally manageable. Finally, this book showed that the results from the simple schema theories provided the inspiration for the experiments performed with the Markov model of an EA, thus indicating that a theory need not be totally predictive to be useful.

A final concern is with the traditional, empirical methodology often used in the EA community, in which an EA is carefully tuned so that it outperforms some other algorithm on a few ad hoc problems (e.g., the De Jong 1975 test suite). Unfortunately, the results of such studies typically have only weak predictive value regarding the performance on new problems.

Interestingly, this form of empirical methodology was also used until recently in the concept learning portion of the machine learning community. Various concept learners were extensively optimized to perform well on small sets of concepts. However, their performance on new, previously unseen concepts was generally hard to predict. In response, researchers began focusing on concept generators, which could produce random concept learning problems within a certain class (e.g., Rendell 1990). The key was to identify characteristics of concepts (noise, number of irrelevant attributes, etc.) that affected different learning algorithms in different ways. The goal was to match concept learning algorithms with specific concept characteristics. This approach yielded results that were more informative and predictive.

This book suggests that the EA community follow the same trend, first by finding important problem characteristics that affect the performance of EAs, and then by creating test-problem generators in which those characteristics can be methodically changed. This book gave one example of this process, by theoretically illustrating the importance of multimodality, and then by using this insight to create a multimodality test-problem generator. Certainly, the multimodality generator introduced here is not the first such generator, nor will it be the last. But the concept of problem generators is one that deserves emphasis, and it is hoped that EA researchers will focus more energy not just on improving algorithms, but also on finding appropriate problems for testing those algorithms.

On a related issue, it is common practice to run EAs to some fixed termination criteria, and then to report the results only after termination. However, this ignores the dynamic aspects of an EA, and can lead to overly general conclusions. For example, as was seen in Chap. 14, conclusions can often turn out to be surprisingly dependent on the termination criteria, often reversing if a different cutoff is used. This has also been noted by Segre et al. (1991) in the explanation-based learning community. From both an engineering and scientific standpoint, it is crucial to include results throughout the running of EAs. Thus, this book suggests that a second way to improve empirical methodology is to always show results over the whole running time of an EA.

Appendix

Formal Computations for Aggregation

This appendix formally computes $r_{\{i \vee j\},k}$. Let S_t be the random variable for the Markov chain, which can take on any of the N state values at time t. Then the short-hand notation $p_{i,j}$ is really $P(S_t = j \mid S_{t-1} = i)$ and ${p_i}^{(t)}$ is really $P(S_t = i)$. Recall the definition of conditional probability: $P(A \mid B) = P(A \wedge B)/P(B)$. Recall also the definition for "averaging" probabilities: $P(A) = \sum_l P(A \wedge B_l)$ where the B_l's are mutually exclusive and exhaust the space. The computation of $r_{\{i \vee j\},k}$ is straightforward. By definition:

$$r_{\{i \vee j\},k} = P(S_t = k \mid S_{t-1} = (i \vee j))$$

By definition of conditional probability and by expanding the disjunctions:

$$r_{\{i \vee j\},k} = \frac{P(S_t = k \wedge S_{t-1} = (i \vee j))}{P(S_{t-1} = (i \vee j))}$$

$$r_{\{i \vee j\},k} = \frac{P(S_t = k \wedge S_{t-1} = i) + P(S_t = k \wedge S_{t-1} = j)}{P(S_{t-1} = i) + P(S_{t-1} = j)}$$

Expanding via the "averaging" of probabilities yields:

$$r_{\{i \vee j\},k} = \frac{\sum_l P(S_t = k \wedge S_{t-1} = i \wedge S_{t-2} = l) + \sum_l P(S_t = k \wedge S_{t-1} = j \wedge S_{t-2} = l)}{\sum_l P(S_{t-1} = i \wedge S_{t-2} = l) + \sum_l P(S_{t-1} = j \wedge S_{t-2} = l)}$$

Using the definition of conditional probability several times, and the fact that the process is Markovian yields (in short-hand notation):

$$r_{\{i \vee j\},k} = \frac{p_{i,k} \sum_l p_{l,i} \; {p_l}^{(t-2)} + p_{j,k} \sum_l p_{l,j} \; {p_l}^{(t-2)}}{\sum_l p_{l,i} \; {p_l}^{(t-2)} + \sum_l p_{l,j} \; {p_l}^{(t-2)}}$$

What is interesting to note is the time-dependence of this expression. Since the ${p_l}^{(t-2)}$ values are not known in advance, one can only make an assumption of "uniformity" (i.e., that the ${p_l}^{(t-2)}$ values are the same for all l). If this is done the time-independent expression obtained is:

$$r_{\{i \vee j\},k} = \frac{m_i \, p_{i,k} + m_j \, p_{j,k}}{m_i + m_j}$$

where m_i and m_j are the sums of the probability mass in columns i and j. This is what was obtained more intuitively in Chap. 13.

Now clearly the uniformity assumption will be wrong in general, which explains why the averaging procedure can lead to errors in numerical computations. However, under conditions of row or column equivalence it is trivial to show that both the time-dependent and time-independent forms lead to the same time-independent answers. Thus, under row or column equivalence the uniformity assumption is irrelevant, and the averaging procedure yields no error. Under row and column similarity the uniformity assumption is nearly irrelevant and the time-independent expression is a good approximation for the time-dependent expression. The error of this approximation is computed in Chap. 13.

References

Bäck, T. and H.-P. Schwefel (1993). An overview of evolutionary algorithms for parameter optimization. *Evolutionary Computation 1*(1), 1–23.

Belew, R. and L. Booker (Eds.) (1991). *International Conference on Genetic Algorithms*, Volume 4, *ICGA91*. Morgan Kaufmann, San Mateo, CA.

Booker, L. (1992). Recombination distributions for genetic algorithms. In Whitley 1992 (Ed.), *FOGA92*, pp. 29–44.

Box, G. E. P. (1957). Evolutionary operation: A method of increasing industrial productivity. *Applied Statistics 6*, 81–101.

Christiansen, F. (1989). The effect of population subdivision on multiple loci without selection. In M. Feldman (Ed.), *Mathematical Evolutionary Theory*, pp. 71–85. Princeton University Press, Princeton, NJ.

Dasgupta, D. and Z. Michalewicz (Eds.) (1997). *Evolutionary Algorithms in Engineering Applications*. Springer-Verlag, Berlin.

Davidor, Y. (1990). Epistasis variance: A viewpoint of GA-hardness. In G. Rawlins (Ed.), *Foundations of Genetic Algorithms*, Volume 1, pp. 23–35. Morgan Kaufmann, San Mateo, CA.

Davis, L. (1989). Adapting operator probabilities in genetic algorithms. In Schaffer 1989 (Ed.), *ICGA89*, pp. 60–69.

Davis, T. and J. Principe (1991). A simulated annealing like convergence theory for the simple genetic algorithm. In Belew and Booker 1991 (Ed.), *ICGA91*, pp. 174–181.

Dayar, T. and W. Stewart (1997). Quasi-lumpability, lower bounding coupling matrices, and nearly completely decomposable Markov chains. *SIAM Journal Matrix Analysis and Applications 18*(2), 482–498.

Deb, K. and D. Goldberg (1992). Analyzing deception in trap functions. In Whitley 1992 (Ed.), *FOGA92*, pp. 93–108.

De Jong, K. (1975). *Analysis of the Behavior of a Class of Genetic Adaptive Systems*. Ph. D. thesis, University of Michigan, Ann Arbor, MI.

De Jong, K. (1985). Genetic algorithms: A 10 year perspective. In J. Grefenstette (Ed.), *International Conference on Genetic Algorithms*, Volume 1, pp. 169–177. Carnegie Mellon University, Pittsburgh, PA.

De Jong, K. and W. Spears (1989). Using genetic algorithms to solve NP-complete problems. In Schaffer 1989 (Ed.), *ICGA89*, pp. 124–132.

De Jong, K., M. Potter, and W. Spears (1997). Using problem generators to explore the effects of epistasis. In T. Bäck (Ed.), *International Conference on Genetic Algorithms*, Volume 7, pp. 338–345. Morgan Kaufmann, San Francisco, CA.

De Jong, K., W. Spears, and D. Gordon (1994). Using Markov chains to analyze GAFOs. In Whitley and Vose 1994 (Ed.), *FOGA94*, pp. 115–137.

Eshelman, L., R. Caruana, and D. Schaffer (1989). Biases in the crossover landscape. In Schaffer 1989 (Ed.), *ICGA89*, pp. 10–19.

Eshelman, L. and D. Schaffer (1991). Preventing premature convergence in genetic algorithms by preventing incest. In Belew and Booker 1991 (Ed.), *ICGA91*, pp. 115–122.

Fogel, D. (1992). *Evolving Artificial Intelligence*. Ph. D. thesis, University of California, San Diego, CA.

Fogel, D. B. (1995). *Evolutionary Computation*. IEEE Press, New York.

Fogel, L., A. Owens, and M. Walsh (1966). *Artificial Intelligence Through Simulated Evolution*. Wiley, New York.

Fraser, A. (1957). Simulation of genetic systems by automatic digital computers I: Introduction. *Australian Journal of Biological Science 10*, 484–491.

Fukuda, T., K. Mori, and M. Tsukiyama (1999). Parallel search for multi-modal function optimization with diversity and learning of immune algorithm. In D. Dasgupta (Ed.), *Artificial Immune Systems and their Applications*, pp. 211–220. Springer-Verlag.

Garey, M. and D. Johnson (1979). *Computers and Intractability: A Guide to the Theory of NP-Completeness*. Freeman, San Francisco, CA.

Geiringer, H. (1944). On the probability theory of linkage in Mendelian heredity. *Annals of Mathematical Statistics 15*, 25–57.

Giordana, A. and F. Neri (1995). Search-intensive concept induction. *Evolutionary Computation 3*(4), 375–416.

Goldberg, D. (1987). Simple genetic algorithms and the minimal, deceptive problem. In L. Davis (Ed.), *Genetic Algorithms and Simulated Annealing*, pp. 74–88. Morgan Kaufmann, San Mateo, CA.

Goldberg, D. (1991). Don't worry, be messy. In Belew and Booker 1991 (Ed.), *ICGA91*, pp. 24–30.

Goldberg, D. E. (1989). *Genetic Algorithms in Search, Optimization, and Machine Learning*. Addison-Wesley, Reading. MA.

Goldberg, D., K. Deb, and J. Horn (1992). Massive multimodality, deception, and genetic algorithms. In R. Männer and B. Manderick (Eds.), *Parallel Problem Solving from Nature*, pp. 37–46. North-Holland, Amsterdam.

Goldberg, D. and J. Richardson (1987). Genetic algorithms with sharing for multimodal function optimization. In Grefenstette 1987 (Ed.), *ICGA87*, pp. 41–49.

Goldberg, D. and D. Segrest (1987). Finite Markov chain analysis of genetic algorithms. In Grefenstette 1987 (Ed.), *ICGA87*, pp. 1–8.

Grefenstette, J. (Ed.) (1987). *International Conference on Genetic Algorithms*, Volume 2, *ICGA87*. Lawrence Erlbaum, Hillsdale, NJ.

Grefenstette, J. (1989). A system for learning control strategies with genetic algorithms. In Schaffer 1989 (Ed.), *ICGA89*, pp. 183–190.

Hart, W. and R. Belew (1991). Optimizing an arbitrary function is hard for the genetic algorithm. In Belew and Booker 1991 (Ed.), *ICGA91*, pp. 190–195.

Holland, J. (1975). *Adaptation in Natural and Artificial Systems*. University of Michigan Press, Ann Arbor, MI.

Holland, J. (1986). Escaping brittleness: The possibilities of general-purpose learning algorithms applied to parallel rule-based systems. In R. Michalski, J. Carbonell, and T. Mitchell (Eds.), *Machine Learning: An Artificial Intelligence Approach*, Volume 2, pp. 593–624. Morgan Kaufmann, San Mateo, CA.

Howe, E. and C. Johnson (1989a). Aggregation of Markov processes: Axiomatization. *Journal of Theoretical Probability 2*(2), 201–208.

Howe, E. and C. Johnson (1989b). Linear aggregation of input-output models. *SIAM Journal Matrix Analysis and Applications 10*(1), 65–79.

Jones, T. (1995). *Evolutionary Algorithms, Fitness Landscapes, and Search*. Ph. D. thesis, University of New Mexico, Albuquerque, NM.

Kemeny, J. and J. Snell (1960). *Finite Markov Chains*. Van Nostrand, New York.

Kennedy, J. and W. Spears (1998). Matching algorithms to problems: An experimental test of the particle swarm and some genetic algorithms on the multimodal problem generator. In *IEEE International Conference on Evolutionary Computation*, pp. 78–83. IEEE Press, New York.

Koza, J. (1992). *Genetic Programming: On the Programming of Computers by means of Natural Selection.* MIT Press, Cambridge, MA.

Koza, J., F. Bennett, D. Andre, and M. Keane (1999). *Genetic Programming III: Darwinian Invention and Problem Solving.* Morgan Kaufmann, San Francisco, CA.

Mahfoud, S. (1995). *Niching Methods for Genetic Algorithms.* Ph. D. thesis, University of Illinois at Urbana-Champaign.

Michalewicz, Z. (1999). *Genetic Algorithms + Data Structures = Evolution Programs.* Springer-Verlag, Berlin.

Michalewicz, Z. and D. Fogel (2000). *How to Solve it: Modern Heuristics.* Springer-Verlag, Berlin.

Miller, R. G. (1986). *Beyond ANOVA, Basics of Applied Statistics.* Wiley, New York.

Mitchell, D., B. Selman, and H. Levesque (1992). Hard and easy distributions of SAT problems. In *Proceedings of the Tenth National Conference on Artificial Intelligence*, pp. 459–465. AAAI Press/MIT Press, Cambridge, MA.

Moon, B. and T. Bui (1994). Analyzing hyperplane synthesis in genetic algorithms using clustered schemata. In Y. Davidor, H.-P. Schwefel, and R. Männer (Eds.), *Parallel Problem Solving from Nature Conference, Lecture Notes in Computer Science*, Volume 866, pp. 108–118. Springer-Verlag, Berlin.

Muller, H. J. (1964). The relation of recombination to mutational advantage. *Mutat. Res. 1*, 2–9.

Mühlenbein, H. (1998). The equation for response to selection and its use for prediction. *Evolutionary Computation 5*(3), 303–346.

Nix, A. and M. Vose (1992). Modeling genetic algorithms with Markov chains. *Annals of Mathematics and Artificial Intelligence 5*, 79–88.

O'Reilly, U.-M. and F. Oppacher (1994). The troubling aspects of a building block hypothesis for genetic programming. In Whitley and Vose 1994 (Ed.), *FOGA94*, pp. 73–88.

Rechenberg, I. (1973). *Evolutionsstrategie: Optimierung technischer Systeme nach Prinzipien der biologischen Evolution.* Frommann-Holzboog, Stuttgart.

Rees, J. and G. Koehler (1999). An investigation of GA performance results for different cardinality alphabets. In L. Davis, K. De Jong, M. Vose, and D. Whitley (Eds.), *Evolutionary Algorithms*, Volume 111 of *IMA Volumes in Mathematics and its Applications*, pp. 191–206. Springer-Verlag.

Rendell, L. (1990). Feature construction for concept learning. In D. Benjamin (Ed.), *Change of Representation and Inductive Bias*, pp. 327–353. Kluwer, Boston, MA.

Robbins, R. (1918). Some applications of mathematics to breeding problems, III. *Genetics 3*, 375–389.

Rudolph, G. (1993). Massively parallel simulated annealing and its relation to evolutionary algorithms. *Evolutionary Computation 1*(4), 361–383.

Ryan, C. (1995). Racial harmony and function optimization in genetic algorithms – the races genetic algorithm. In *Proceedings of the Evolutionary Programming Conference*, Volume 4. MIT Press, Cambridge, MA.

Schaffer, D. (Ed.) (1989). *International Conference on Genetic Algorithms*, Volume 3, *ICGA89*. Morgan Kaufmann, San Mateo, CA.

Schaffer, D. and L. Eshelman (1991). On crossover as an evolutionarily viable strategy. In Belew and Booker 1991 (Ed.), *ICGA91*, pp. 61–68.

Schwefel, H.-P. (1981). *Numerical Optimization of Computer Models.* Wiley, Chichester, UK.
Schwefel, H.-P. (1995). *Evolution and Optimum Seeking.* Wiley, Chichester, UK.
Segre, A., C. Elkan, and A. Russell (1991). A critical look at experimental evaluations of EBL. *Machine Learning 6*(2), 183–195.
Smith, G. (1979). Adaptive Genetic Algorithms and the Boolean Satisfiability Problem. Technical report, University of Pittsburgh, Pittsburgh, PA.
Smith, R., S. Forrest, and A. Perelson (1992). Population diversity in an immune system model: Implications for genetic search. In Whitley 1992 (Ed.), *FOGA92*, pp. 153–165.
Smith, S. (1983). Flexible learning of problem solving heuristics through adaptive search. In *Joint Conference on Artificial Intelligence*, Volume 8, pp. 422–425. William Kaufmann.
Spears, W. (1994). Simple subpopulation schemes. In A. Sebald and L. Fogel (Eds.), *Proceedings of the Evolutionary Programming Conference*, Volume 3, pp. 296–307. World Scientific, Singapore.
Spears, W. (1999). Aggregating models of evolutionary algorithms. In P. Angeline (Ed.), *Congress on Evolutionary Computation*, pp. 631–638. IEEE Press, New York.
Spears, W., K. De Jong, T. Bäck, D. Fogel, and H. de Garis (1993). An overview of evolutionary computation. In P. Brazdil (Ed.), *European Conference on Machine Learning, Lecture Notes in Artificial Intelligence*, Volume 667, pp. 442–459. Springer-Verlag, Berlin.
Spears, W. and K. De Jong (1996). Analyzing GAs using Markov models with semantically ordered and lumped states. In D. Whitley and M. Vose (Eds.), *Foundations of Genetic Algorithms*, Volume 4, pp. 85–100. Morgan Kaufmann, San Francisco, CA.
Stark, H. and J. Woods (1986). *Probability, Random Processes, and Estimation Theory for Engineers.* Prentice Hall, Englewood Cliffs, NJ.
Stewart, W. (1994). *Introduction to the Numerical Solution of Markov Chains.* Princeton University Press, Princeton, NJ.
Stewart, W. and W. Wu (1992). Numerical experiments with iteration and aggregation for Markov chains. *ORSA Journal on Computing 4*(3), 336–350.
Suzuki, J. (1993). A Markov chain analysis on a genetic algorithm. In S. Forrest (Ed.), *International Conference on Genetic Algorithms*, Volume 5, pp. 146–153. Morgan Kaufmann, San Mateo, CA.
Syswerda, G. (1989). Uniform crossover in genetic algorithms. In Schaffer 1989 (Ed.), *ICGA89*, pp. 2–9.
van Nimwegen, E., J. Crutchfield, and M. Mitchell (1997). Finite populations induce metastability in evolutionary search. *Physics Letters A 229*, 144–150.
Vose, M. (1992). Modeling simple genetic algorithms. In Whitley 1992 (Ed.), *FOGA92*, pp. 63–74.
Vose, M. (1995). Modeling simple genetic algorithms. *Evolutionary Computation 3*(4), 453–472.
Whitley, D. (1989). The GENITOR algorithm and selection pressure: Why rank-based allocation of reproductive trials is best. In Schaffer 1989 (Ed.), *ICGA89*, pp. 116–121.
Whitley, D. (1992). An executable model of a simple genetic algorithm. In Whitley 1992 (Ed.), *FOGA92*, pp. 45–62.
Whitley, D. (Ed.) (1992). *Foundations of Genetic Algorithms*, Volume 2, *FOGA92*. Morgan Kaufmann, San Mateo, CA.
Whitley, D. and M. Vose (Eds.) (1994). *Foundations of Genetic Algorithms*, Volume 3, *FOGA94*. Morgan Kaufmann, San Francisco, CA.

Winston, W. (1991). *Operations Research: Applications and Algorithms.* PWS-Kent, Boston, MA.

Wolpert, D. and W. Macready (1995). No Free Lunch theorems for search. Technical Report 95-02-010, Santa Fe Institute, Santa Fe, NM.

Wolpert, D. and W. Macready (1997). No Free Lunch theorems for optimization. *IEEE Trans. on Evolutionary Computation 1*(1), 67–82.

Index